Science: Growth and Change

HENRY W. MENARD

Science: Growth and Change

Harvard University Press   Cambridge, Massachusetts   1971

Q
125
.M437
1971

© Copyright 1971 by the President and Fellows of Harvard College
All rights reserved
Distributed in Great Britain by Oxford University Press, London
Library of Congress Catalog Card Number 77-156138
SBN 674-79280-7
Printed in the United States of America

To Gifford

whose wifely eyes have also seen
five doubling periods in oceanography
and the changes that accompanied them

Preface

This study has grown as aimlessly and at as variable rates as the science that it describes. It began in 1965 when I spent a year in the Office of Science and Technology, which is the White House science staff. I had the pleasure of first looking into Price's *Little science, big science* in the library of the Bureau of the Budget, and the present volume is an outcome. I was at that time responsible for trying to understand the growth and prospects for oceanography, a science in which I had been engrossed for sixteen years. I began to apply Price's techniques of analysis to Federal funding for oceanography and such other things as the quality of research at different oceanographic laboratories and the consequences of the extremely rapid growth of the science. Oceanography, it developed, doubled every four to five years, and it was immediately clear that this had had remarkably intense effects and that it could not long continue.

When again in La Jolla, I introduced my students to the history and sociology of marine geology. They seemed to be fascinated, especially because they could see the effects of rapid growth in everything that they were doing. Questions began to arise concerning the growth of other sciences, and it was only natural to examine the sociology of geology, the parent of marine geology. There the growth rates in some fields proved to be very slow, and they seemed to account for some sociological effects that had been unexplained and even unremarked.

More and more examples accumulated of highly variable growth rates and of their influence on scientific careers. Gradually the study began to incorporate calculations of the effects of variable growth on a wide range of scientific concerns: studying, teaching, research, publishing, citations, the basis of scientific prestige, promotions, unemployment. It became apparent that generally unsuspected forces have a powerful influence on the careers of scientists and everyone else enmeshed in rapid change.

This book presents some evidence for change and speculates about many effects. It also makes a beginning toward understanding the forces that cause and do not cause change. It may provide some solace for those in dormant fields who have wondered why life has passed them by. It may, perhaps, make a few highly successful scientists a little more modest. Most of all it may guide those who still have a choice—and so have we all.

H. W. MENARD

Institute of Marine Resources
and Scripps Institution
of Oceanography,
University of California, San Diego

Contents

	Preface	vii
1	Introduction	1
2	Measuring the Growth of Literature	17
3	Growth of Sciences	41
4	Population and Other Factors Affecting Growth	59
5	Papers and Citations and Scientific Fame	84
6	Scientific Literature	129
7	Education	157
8	A Department of Science	185
9	Scientists in Society	194
	Index	209

Tables

4.1	Census of Some Subfields in Science	61
4.2	Activities of Scientists	68
4.3	Employment of Scientists	69
4.4	Population of American Physicists Compared with World Output of Papers in Physics and Its Subfields	73
5.1	Prolific Earth Scientists	88
5.2	Publication Records of Some Members of the National Academy of Sciences in Different Fields	94
5.3	Productivity of Papers by Some Members of the National Academy of Sciences Compared with Citations in the *Science Citation Index* for 1968	95
6.1	Distribution of Citations in Papers in Earth Sciences at Different Times Compared with an Average Distribution for Science as a Whole	137
7.1	Number of Doctoral Degrees Granted per Doctorate-holding Staff Member in Science and Engineering in 1961 for All Activities and Segments in Colleges and Universities, Tabulated by Academic Field	174

Figures

1.1	The scientific viewpoint	*facing page* 1
1.2	The growth of American scientists as a proportion of the total population	8
1.3	Growth of different types of scientific information	15
2.1	Age of earth scientists at election to the National Academy of Sciences	23
2.2	Comparison of indexed papers and citation ages in marine geology	27
2.3	Annual production of indexed papers in marine geology compared with number of citations	28
2.4	Comparison of cumulative papers in nuclear astrophysics with age of citations	29
2.5	Papers and citations in nuclear astrophysics	30
2.6	Growth of papers in vertebrate paleontology for four centuries	32
2.7	Papers and citations in vertebrate paleontology	32
2.8	Growth of papers in glacial geology	34
2.9	Papers and citations in glacial geology and geomorphology	35
2.10	Age of citations in individual selected papers in the *Astronomical Journal*	37
2.11	Age of citations in individual papers in *Bulletin of the Geological Society of America*	37
3.1	Annual production of abstracted or indexed papers in various sciences	41
3.2	Pages of various components of American geological literature	43
3.3	Citation age distributions in the *Bulletin of the Geological Society of America*	45
3.4	Agglomeration of citations in different botanical journals	46
3.5	Papers and citations in several lively transitional fields	48
3.6	Output of papers in some subfields of physics	51
3.7	Growth of stable and cyclical subfields in the earth sciences	56
3.8	Subfields with rapid growth	56
4.1	Science and technology components of the general population and labor force	59
4.2	A historical census of American scientists with some specialties differentiated	60
4.3	Growth of the population of geologists and graduates with geology majors	63

4.4	Comparison of number of geologists and number of pages of geological reports	71
4.5	Federal funding for research and development by fiscal years	76
4.6	Federal funding for components of research and development	79
4.7	Federal expenditures compared with output of scientific papers and doctorates	80
4.8	Federal funding for basic research compared with output of science	81
5.1	The scientific literature	86
5.2	The decline with age in the fraction of papers by NAS members that are cited in 1968	97
5.3	Comparison of published and cited papers by NAS members	98
5.4	The fraction of authors writing N papers who are cited C times in the literature of geology	99
5.5	The fraction of authors writing N papers who have their first paper cited in the *Bulletin of the Geological Society of America*	101
5.6	The distribution of citations to the publications of R. A. Daly	105
5.7	The number of citations, C, received by papers of R. A. Daly in 5-year intervals after publication	108
5.8	The number of authors who write at least N papers plotted as a function of N	110
5.9	The number of citations, C, received by papers of various prolific authors in slowly expanding specialties	112
5.10	The distribution of citations to my publications	114
5.11	The number of citations, C, received by my papers in marine geology	117
5.12	Citations to a revolutionary paper, "History of ocean basins"	121
6.1	The emphasis on formal bibliography during the dormant period in the earth sciences	135
6.2	Distribution of citations in earth science literature normalized to the same publication date	139
6.3	The growth of various components of the Geological Survey	149
6.4	The cumulative cost and output in pages of all state geological surveys	154
7.1	Educational staff and output in the earth sciences	160
7.2	Doctorates in sciences and engineering	163
7.3	The number of professors increases without regard for the number of graduates in the earth sciences	167
7.4	Growth of doctorates and papers in subfields of the earth sciences	168
7.5	Growth of doctorates in subfields plotted for comparison	169
7.6	Relationship between doctorates in subfields and the number of college departments in earth sciences	171
7.7	Growth of doctorates in subfields in geology normalized to the same starting time	173
7.8	Growth of subfields plotted as if total of papers in each were the same at present	181
8.1	Annual budgets of federal agencies concerned with the environmental sciences compared with the nondefense and total federal budget	188
8.2	During the nineteenth century each new agency was superimposed on a relatively stable older one	189
8.3	Budgets of environmental agencies normalized to the same founding date and starting budget	190

Science: Growth and Change

Fig. 1.1 The scientific viewpoint. Reprinted from C. E. Wegmann, 1939, "Zwei Bilder für das Arbeitszimmer eines Geologen," *Geologische Rundschau 30*, 391.

1 Introduction

> You'll get no promotion this side of the ocean,
> So cheer up, my lads, bless 'em all!

This song from World War II displays the common understanding that promotions and prosperity are not uniformly distributed even among the uniformed.[1] It is also dead wrong because of a misunderstanding of the factors involved. So is it elsewhere in life, among the high and the low, the ignorant and the educated, the student, the teacher, and the scientist. The effects of growth rates are little known, and, for lack of a little analysis and counseling, people embark on careers that are virtually doomed from the start. Skilled workers tie themselves to maximum proficiency with a lathe or a lunar lander and wonder what hit them when the tool becomes obsolete. Growth, change, stability, contraction, decay are all around us, and they affect all aspects of our lives. In this book we examine the narrower subject of how they affect the careers of scientists.

COMMON EXPERIENCES WITH GROWTH RATES

First, let us return to the soldier going overseas and his chances of promotion. We can visualize the size of the army as constant, expanding, or contracting, and clearly the probability of promotion will vary according to which state prevails. The song assumes a steady state in which the slackers remaining in the States will have no way to go upward. Meanwhile, the heroes in the trenches will die like flies, and the sole survivor of the dawn patrol will be brevetted

---

[1] Jimmy Hughes, Frank Lake, and Al Stillman, 1941, "Bless 'em all" (New York, Sam Fox Publishing Co., Inc.).

from private to master sergeant. Thus it was in World War I, when the British and German troops went straight from the training fields to the trenches. Wars as idiotically led as the First World War are rare, however, because the troops naturally revolt, as did the French. This is hardly a normal state and is not the one to anticipate if you want a successful military career. Among other things, the career might be brief.

The thing to do is stay home as long as possible. This is a difficult course of action for a soldier when the cannon thunder, but consider the consequences. I met a second lieutenant in the Americal division in New Caledonia in 1942. He complained that he was unpromoted although his division had been overseas longer than any other in our army. Meanwhile, all his classmates in the States were captains and majors and were marrying the prettiest girls left behind. He wondered why he was the victim of injustice; but he was not — he was merely uninformed. How could he be promoted when the table of organization was full? His friends were manning the new divisions then forming, and as the expansion continued, a few months as a captain qualified them for major. In the air force it might qualify them for colonel, and indeed jokes appeared about colonels too young to buy a drink legally. But who else should be a colonel in the air force as it expanded so rapidly? There were not enough old majors to meet the demand. We need merely look at the promotion records of our most eminent generals to see how they jumped through four or five ranks as the Army of the United States grew at home.

After each war the armed forces contract, and the possibilities for promotion are slim. Even after disposing of reserves and draftees, every rank is overflowing with combat veterans who are not merely young but very young for the rank. Fortunately, this problem is dealt with by constant evaluations, fitness reports, physical examinations, and a resultant weeding by early retirement. The general quality and capacity of the armed forces are thus enhanced, because all the less competent men can be eliminated. The problems of promotions and morale are solved if someone has the foresight to shrink the forces to a size from which they must surely expand. Alternatively, yet another war may produce expansion.

Now and then the army exists in a steady state for a period long enough to observe. In the United States such a condition prevailed between the world wars. Officers felt fortunate if promotions came

Introduction

once a decade, and a wife with money was a big help. It speaks well for the loyalty and also the organization of the army that it continued to exist without attempting a revolt. Somehow the aging officers were thinned out and the remainder kept busy. We need only look elsewhere to see what might have happened. Generals at the top of a successful career rarely lead revolutions. It is the unknown majors and colonels, stymied in seeking advancement, who suddenly appear on the palace balcony.

The song of the great depression, "Brother can you spare a dime?," states the impact on an individual of a change in economic growth. It is but another example of the same influences that affect the military. A young man who went to work in the computer business twenty years ago had a different song to sing: "How to succeed in business without really trying." The business expanded, and everyone was promoted because even a dubiously competent man could hardly guess wrong, and anyway he knew more than a total stranger. Mistakes in administration were readily forgiven, and more likely they were not even perceptible as profits skyrocketed. In such a climate a good young engineer who was not advancing fast enough to suit himself could always found his own company. It, in turn, also prospered and at an even faster rate than the parent company. Soon expansion began, and the advent of the second factory was accompanied by a sudden influx of administrators not previously seen. The expansion also required money, which brought in the banks and venture capitalists, and the founder no longer owned his business. What was once fun for a small group became work for a large one, and new engineers split off and founded new companies. The accent is on youth, just as it was in the air force in 1942 and for the same reason. Indeed, it is the same reason that children dominate any country with a population-growth problem. In any field with rapid expansion, almost everyone is newly arrived.

How different it is in a business, or any other organization, which is in a steady state. "Steady state" in a business means that it is growing at the same rate as the gross national product or the population or some similar measure. We have available an aphorism indicating yet another widespread realization of the effects of growth rates: "If you are standing still, you are falling behind." A utility company is roughly in a steady state, although it may grow somewhat faster than the population because of increasing per capita use. In such a company the employees live in an atmosphere of stability, steady and growing prosperity, and aging. The matter is complicated by turnover, but in general the increment of new permanent employees is small each year by the very definition of a stable industry. Thus, twenty- and thirty-year men are common, and the company magazine has lists of recipients of gold watches inscribed "For long and faithful service." The atmosphere is conservative, promotion is from within the company, and labor problems are minimal.

Once again, how different it is in a declining business such as the manufacture of buggy whips, transport aircraft, or DDT. Once again, there are no more promotions, and soon the cuts begin. We are now well into this stage in the giant aerospace industry. Tens of thousands of employees including thousands of engineers are out of work or are about to be. The salaries of principal officials are being reduced; the trust funds prudently collected to pay unemployment benefits are exhausted.

The contrast between the buggy whip and aerospace industries is important, because the increased utilization of advancing technology will make the latter typical. The buggy whip or at least the buggy industry was important in its time; it used a variety of components, skilled craftsmen, and so on. Despite its eventual decline, however, its general characteristic was persistence for centuries. Even the decline was gradual, although much faster than the rise. Thus, people had about the same salaries as in other relatively stable industries. They also had abilities which could be put to other purposes except for a small fraction of specialists.

In the aerospace industry the engineers are highly paid compared with stable industries, because they were in short supply during expansion. The large majority have become very specialized technically, and others have become youthful managers of specialists. When they lose their jobs, few can find others that will pay them anything like the same salary. The high-paying jobs are in some other specialty. Many of the more enterprising, and these are a youthful and energetic group, try to transfer to oceanography or pollution abatement or some other field that looks like it has a future. Usually this requires that they go back to school, which in turn requires money. Unfortunately, they have been buying and mortgaging on the assumption that the expansion and rapid promotions of the past would continue; they have debts instead of money. They are the victims of a growth cycle

Introduction

that will become ever more common. In an extreme case, the National Aeronautics and Space Agency recently closed its enormous testing facility in a formerly rural county of Mississippi. Most of the employment in the area has vanished, and people are leaving or are on relief. Meanwhile, the county has a large debt for public facilities which it acquired, at NASA's insistence, to meet the demands of the large work force. We hear much at the moment about paying the cost of solid waste disposal by including it in the original purchase price of the new item which will one day be junk. Somehow we need to include the cost of the new schools and sewers before we build a new industry. So in Los Angeles advocates of "zero growth" appear. But new industries must arise as technology changes, and they must be staffed with technical specialists, and soon they will decline. Thus, zero growth or not, provision must be made for the reeducation of all specialized employees as a part of the cost of production. It is apparent that the incremental cost for retraining will vary inversely with the life of the industry. Like many other factors influencing careers and lives, it is a function of growth rate.

THE GROWTH OF SCIENCE

We turn now to our principal concern, the growth of science and its influence on the lives of scientists and the structure of science itself. We turn without a song because, in contrast to war and business, I cannot think of one to quote. It is generally appreciated that scientific research has a major influence on the nation and that there are a lot of scientists. However, the numbers have grown so fast that the scientist is still commonly viewed as someone strange and even fearful, and for him no minstrel raptures swell. Indeed, the scientist rarely appears even in literature except in caricature as Dr. Strangelove. Scientists themselves may tend to foster this attitude, because so many have been taught to hold the view that scientists are rare (Fig. 1.1). The older teachers remember science twenty or thirty years ago, when it was like a fundamentalist religion in size and dedication or like a computer business in the early phase of expansion. Now it is different; the business has succeeded and expanded beyond its wildest dreams. In the process, however, it has undergone the usual changes. The proportion of administrators has increased, necessarily increased, and in order to acquire the venture capital for expansion, the control of science has been transferred from scientists themselves to people who want to use it. In this case they are politicians and not bankers, because science is supported in large part by the public purse. This seems only reasonable. In any event, it is an almost inevitable consequence of astonishingly rapid and sustained growth.

Science has grown like a bean sprout, faster and ever faster in a simple exponential expansion for 200 years. Now, like a mature bean plant, the growth is beginning to wane, and science and the life of scientists may never be the same. The growth can be measured most easily by the increase in the number of scientific journals, all of which are catalogued and available in libraries.[2] This measure in the past has been representative of the whole of science, because the journals are a central component. Indeed, "science" can be defined quite reasonably as the material that is published in these journals. A journal typically contains about 10 articles per issue and thus about 120 per year. Each article is written by a scientist, and a typical rate for productive scientists is 3 per year. Thus, one journal is equivalent to the output of 40 scientists, and the number of journals has been equatable to the number of scientists. Whether this is still true and will be in the future depends on whether we accept a definition of a scientist as someone who writes 3 papers a year. Otherwise, the population of scientists becomes only loosely related to the science in journals. Price has shown that about 6 million scientific papers have been published since they were invented in 1665. After an initial period of a century, growth was exponential with a doubling time of about 15 years. The characteristic of exponential growth is that the fractional increase each year is constant regardless of the size of the growing population. Interest compounded at a rate of between 4 percent and 5 percent doubles in 15 years. This growth rate also applies to the number of college graduates and members in scientific institutes, and it grossly approximates the growth in scientific and engineering manpower, although these last figures are from census data and are more difficult to assess.

This growth rate is strikingly faster than that of the United States

---

[2] The nature of the growth has been admirably expounded in two fascinating books by Derek J. de Solla Price: *Science since Babylon,* 1961 (New Haven, Yale University Press) and *Little science, big science,* 1963 (New York, Columbia University Press). My debt to Professor Price is evident throughout this book.

Introduction 7

population as a whole, which doubles about every 50 years or at a rate of little more than 1 percent per year compounded. The difference has surprisingly large effects. For example, banks pay compound interest rates of 4 to 5 percent, but bank accounts do not seem to buy twice as much in 15 years. Remove only 3 to 4 percent for taxes and inflation, however, and the doubling time changes to more than 50 years. One large effect is what Price calls the "immediacy" of science; most science is going on right now. The calculation is straightforward. Every 15 years on average the total number of scientists doubles. The career of a scientist lasts, to take a convenient number, about 45 years. At the end of his career, of each eight scientists who have ever lived, some four will have started in science 15 years earlier. Of the remaining four, two started 30 years before, and of the last two only one was active 45 years earlier. Thus, seven-eighths or 87½ percent of all scientists who ever lived are alive now, and the same fraction of those who lived up to any given time were alive at that time.

In sum, it appears that for centuries the average productivity of scientists remained constant despite an enormous increase in their number. During that time, however, the proportion of scientists in the general population steadily increased. This relationship clearly is limited, because in less than another century everyone would be a scientist. This fact caused Price to predict in the early 1960's that the growth of science would soon slow. This is now confirmed in the sense that the funds for federal support of research and development have remained constant at 3 percent of the gross national product for eight years (Fig. 1.2).[3] The budget of the National Science Foundation once grew very rapidly, but for some years it has been cut by shocking amounts. The support of research by other government agencies is also level or actually dropping, and it appears that support for *basic* science is not even remaining a constant fraction of either the GNP or the federal budget. The effects on manpower are more difficult to evaluate because of inherent delays in making students aware of or even interested in changes in job opportunities. However, good graduate students are drawn to generous fellowships — other things being equal, or in some cases even if they are not. For the first time in a decade some graduate students in the best science departments do not have fellowships or jobs as part-time research assist-

[3] J. P. Martino, 1969, "Science and society in equilibrium," *Science 165,* 769–772.

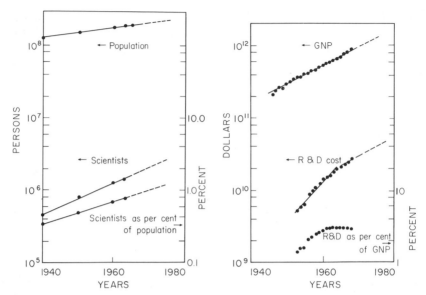

Fig. 1.2 The growth of American scientists as a proportion of the total population; and the stabilization of expenditures for research and development as a constant proportion of the gross national product (after J. P. Martino, 1969 "Science and society in equilibrium," *Science 165*, 770).

ants. This is bound to have some effect on the future population of scientists. Likewise, it is too soon to see whether the number of scientific volumes will continue to grow at the same rate or will wane. Scientists for three centuries have bemoaned the waxing flood of scientific papers and at any time would have liked to see fewer journals published.[4]

The number of scientists is far less than the population, but so is the number of authors who can write three scientific papers per year. It is by no means obvious that "scientists" are not already more numerous than such authors. If so, this average productivity has been declining in recent years and will continue to do so. The effects will be particularly obvious soon if the basic research component of science continues to receive unfavorable treatment. The number of so-called scientists will grow according to Parkinson's Law, but the authors of papers will be less productive than before.

Predictions of impending doom have caused increasing concern

[4] Provided, of course, that space was available for rapid publication of their own contributions.

among the statesmen and bureaucrats of science who are responsible for its vigor to the extent that it is necessary to support national policy. The issue is whether it is possible to justify a continuing rapid expansion of science in terms of national goals. It is clear enough that this justification has not been successful either in the White House or the Congress. Time and again committees of the National Academy of Sciences and panels of the President's Science Advisory Committee have made searching studies of the future development and need for support of different sciences. Time and again they have found the sciences deserving, but the money has not come. Perhaps to a politician it is not very exciting news that scientists want more money for science — any more than that farmers want money for agriculture or sailors for ships.[5]

Instead of worrying about whether a slowing of scientific growth will occur, Martino analyzes some effects which may be expected according to certain assumptions. He considers two possibilities: that the size and growth of science will be limited first as a constant fraction of the GNP, and second as a constant fraction of the population. It now costs $41,000 per year for salary, office, and laboratory support for a scientist in research and development in the United States. Assuming this is constant and the GNP grows at the present rate, the number of scientists can be predicted and the increase used to calculate how many will need to be trained. On these assumptions, the roughly 200,000 science professors in colleges are enough to train the new scientists needed in 1976. On the other hand, if scientists remain a constant 1 percent of the population and the population can be predicted, it appears that there are already 100,000 more science teachers than will be needed to train the small increment of new science students in 1976. These predictions for only a decade from now are enormously at variance with the estimates by the National Science Foundation that an extra 100,000 professors will be needed.[6] The differences merely reflect the startling effects after a few years of what may seem to be small differences in growth rates. They emphasize how important it is to understand the growth of science and of the population of scientists, students, and teachers.

[5] See Warren Weaver, 1959, "Report of the Special Committee," *Science 130*, 1390–1391, for an amusing suggestion to simplify the committee efforts by having them fill in blanks on a two-cent standard report form.
[6] National Science Foundation, 1967, *Science and engineering staff in universities and colleges, 1965–75*, NSF 67-11 (Washington, Government Printing Office).

Martino speculates on some of the corollaries of his calculations. If research and development grow but professors become fewer, the locus of research will shift outside the universities. Alternatively, the character of universities might change with increased research in proportion to teaching. This would be just the opposite of the increased emphasis on teaching advocated in some circles. However, the analysis refers not to what people say they would like but what in fact they are prepared to pay for. Universities lose money on students and make money on research.

Another corollary is that the age distribution of scientists will shift, with the average gradually increasing. Considering the well-known relation between creativity and age, this suggests a disproportionate drop in scientific output. Finally, advancement will be slower and career opportunities fewer. We have already noted that all of these effects are commonly observed in other fields of work and might have been expected in scientific research.

I have emphasized the characteristic growth rate of science for simplicity, but we must now turn to a more complex reality in which different growth rates exist. Price notes that the doubling time is only 10 years for low-grade science, which includes some high-grade development and engineering.[7] Likewise, it is 20 years for very high-quality science. These differences derive from the important fact that a small fraction of the most productive scientists write a disproportionately large fraction of scientific papers. A big producer may be a hundred times more active than a marginal man. This relationship probably exists in all types of human activity — a movie star is *much* more famous than a starlet — but it is obscured by the difficulty of measurement.

In the sciences the quantity of output, if not the quality, can be determined by counting published papers. A quantitative measure of fame is the annual number of citations to each paper, and this can also be counted. The results of these counts are remarkably consistent. The number of authors producing n papers is proportional to $1/n^2$.[8] If 100 authors each write a single paper, only 25 will write 2, and a mere 6 will write 4, and so on. This relationship may also be examined by totaling the number of papers and authors: 165 authors

[7] Price, *Little science,* p. 6.
[8] Price, *Little science,* p. 43.

Introduction

and 586 or more papers, or an average of 3.5 each. The top 2 authors produce 25 percent of the papers; the top 10 percent produce about 50 percent, the intermediate 40 percent produce 40 percent of the papers, and the bottom 50 percent of authors produce about 10 percent of papers. It should be noted that these numbers refer only to authors, not to scientists as a whole, unless we define scientists as authors.

A more restrictive definition is that a scientist is someone whose work is cited in the scientific literature. This definition was once used in an attempt to calculate the efficiency and effectiveness of government science in oceanography. Many federal agencies at the time were trying to plunge into oceanography and proudly reported ever-increasing numbers of oceanographers on their staffs and ever-increasing appropriations for the science. One of the ways to build up a bureaucracy is to demonstrate momentum. The annual oceanography budget of any agency was divided by the annual number of literature citations to oceanographers identified in that agency. The results gave spectacular variations from about $10,000 per citation for some academic oceanography laboratories to more than $1,000,000 per citation for one government laboratory. Most of the federal scientists clearly did not fit the definition of scientists as authors of cited scientific papers. A related study was made at Woods Hole Oceanographic Institution to test the assumption that the people who were not writing papers for publication probably were writing most of the mimeographed cruise reports and interim studies generated by most large laboratories. The result was completely negative; most of the mimeographed reports were written by the same people who did most of the publishing and received most of the citations. If only the most productive scientists enter a new field, it will grow very rapidly compared to the average. Moreover, the residue of low producers left in an old field will have difficulty in making it grow at all. It is commonly observed that very active scientists do shift subfields when they consider the cream skimmed, so we may anticipate quite different growth rates. If subdivisions of fields, why not subdivisions of science? Why not science as a whole?

If we can measure different growth rates of science, then we shall be in a position to evaluate how they affect the careers and lives of scientists.

ON THE PROBLEMS OF A CENSUS OF THINGS THAT CHANGE RAPIDLY

In principle it would not be difficult to take a census of the live human population of the earth, because the time, place, and species are readily defined. Demographers, economists, and the like are used to censuses of this general sort in which the same kind of observation can be repeated at long intervals, and trends of different variables can be compared to find cause and effect. Even so, there are always uncertainties in the definition of the population that set limits on the significance of possible correlations. In the global census, for example, uncertainties would arise because of the definition of "alive" and because of astronauts who are at the boundary of earth gravity. At present, these would be minor, but it is not difficult to conceive of medical technology capable of producing many stages of "alive" and space technology capable of filling the void with people, and governments eager to do both. Then there would be confusion, because people could be counted in more than one category, and as the technology changed new categories would arise for which previous data did not exist. Time series would begin and end at different times, and cause and effect would be hard to evaluate.

Such a confusion exists in making a census of science or scientists. The historical records do not measure the same things as modern ones mainly because the same things did not exist. A century ago the few scientists generally were counted in other categories such as "physician," "clergyman," or "member nobility," or most likely "professor." Half a century ago there were enough "chemists" and "engineers" to be listed in the U.S. Census, but that does not mean that the same people would be listed in the same way today. Some would be technicians and some would be managers and very likely some would be physicists or physiologists. This trend has continued to the present. The normal rapid expansion of science constantly creates new groups that are large enough to deserve listing in a census but have no previous sociological history.

The speed of the expansion raises two other related problems. The best data always exist for the categories of least interest; and it is very easy for an individual to change categories.

In any given science many subfields can be identified for more than a century. Optics and acoustics in physics and paleontology and mineralogy in geology are examples of persistent subfields. Because

Introduction

they have existed so long, they lend themselves to analysis and comparison with social and economic variables. If the conclusions are generalized to all of a science or even worse to all of science, however, they will almost inevitably be erroneous. This is because the persistent subfields are abnormal. Most of the action is in ones that are created, explode into fragments, and are no longer identifiable after a few decades. The mere fact that a subfield is identifiable for a century virtually guarantees that it is dormant — and science that doubles every 15 years is not.

In a census there is usually a category called "other," and in a rapidly expanding subject this is the one that is apt to be most important. By the time a population group is separately listed even in a fine-toothed census, it consists of perhaps a thousand individuals. In ten doubling periods the group grows to a million, by which time it probably has broken into distinct subgroups. This expansion may take two or three decades, during which the census lists the population. The time series is very short. On the other hand, it takes just as long to expand at a constant rate from one individual to a thousand as it does from a thousand to a million. Consequently, the time series can be doubled if the members of the population can be winkled out of the category "other." It can be done, but only by examining other types of data such as bibliographies or society records to see who was doing what at any given time.

This brings us to the second inherent problem in making a census of a rapidly expanding population. Rapid change implies versatility and background strength. These are exactly the characteristics that enable the best scientists or engineers to change from one specialty to another almost at will. Some of the best, like artists, find challenge mainly in such changes. Fritz Zwicky is said to have remarked that a good scientist ought to be able to shift to a new field and make a significant discovery within a year; and he did, and he did. Less versatile people require retraining to shift from specialties, but it can be done. The economic and social advantages of being in a rapidly expanding field tend to suck people in from more stagnant fields and thereby cause rapid population shifts that are difficult to interpret. The same general kinds of problems occur in enumeration of the work of scientists as in counting the scientists themselves. Mostly the work is identified as published scientific papers and most as listed in abstract journals or bibliographies. In a single field, the indexers do not identify the subfield while the first 50 or 100 papers appear. Then they go through a period when they do not know what to call it. For example, in geological indexes, "ocean⎯⎯⎯⎯" was a common category from 1933 to 1960, but vanished in the 1960's within the category "submarine⎯⎯⎯⎯." If the growth continues only a few more doubling periods, the category becomes too large, splits, and tends to be reassigned to or duplicated in older categories.

It appears that this may be happening to science, which has been a novel, expanding, splitting, and distinct activity for hundreds of years. Now it is so widespread that it may be in process of reintegration into normal human activity. If so, the scientific method will achieve the triumph of becoming the human method.

Problems multiply because sciences overlap and the many abstracting and indexing services serve different ends. It once appeared that a new abstract journal is created for each 300 new scientific journals, which themselves increase tenfold every 50 years (Fig. 1.3).[9] If so, the number would have been about 600 in 1965, but in fact it was 1,885 in 1960 and 3,500 in 1966.[10] It seems that abstract journals grow tenfold in only 30 years. Even so, they are getting out of hand, and the increasingly available computers have been put to use. A listing of computer-generated abstracts and journal searches is a good measure of what is happening.[11] Mere data storage and acquisition systems are excluded, and the remainder are those designed to manipulate existing literature. The number increased from 1 in 1950 to almost 200 in 1970, with only 10 years required for a tenfold expansion.

The different growth rates in Figure 1.3 suggest the degree of overlap. Physicists and chemists and biologists and geologists once worked in different fields, and little occasion for duplicate abstract listings existed. However, *Chemical Abstracts* has subdivisions of "Geochemistry" and "Biochemistry," and presumably most papers in chemical physics or biophysics would justifiably be multiple-listed. This problem is becoming ever more acute, because the universal use of computerized data-processing makes bibliographic work easier. A paper titled "Neutron activation analysis of a fossiliferous sediment from

---

[9] Price, *Little science,* p. 9.

[10] H. R. Malinowsky, 1967, *Science and engineering reference sources* (Rochester, N.Y., Libraries Unlimited, Inc.), pp. 14–15.

[11] *Directory of computerized information in science and technology,* ed. Leonard Cohan, 1968 (New York, Science Associates).

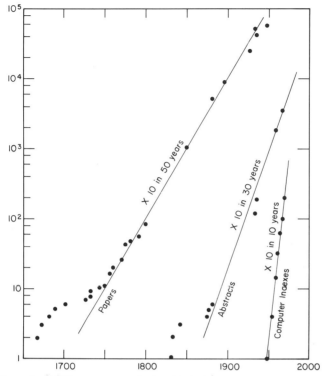

Fig. 1.3 Growth of different types of scientific information.

Mars" ought to appear in almost every abstract journal. Such titles are becoming increasingly common, because of KWIC indexing which puts a premium on long and informative ones. Thus, the apparent growth of science recorded in abstracts and bibliographies may become increasingly inflated and certainly can be misleading.

We can draw the general conclusion that it is much easier to shuffle articles than it is to write them. If we are not careful it may turn out that the information services will become the problem rather than the solution.

The foregoing discussion will serve as an introduction to the analyses of growth patterns and their significance that follow. I use shorter time series than I would like, but this is because of an inherent characteristic of much of the information. For long and detailed studies I use geological examples, because they are the ones most familiar to me. I would prefer more famous or more important examples from other sciences on some points, but it takes a specialist to identify the early parts of a time series for an exploding subject. Likewise, wherever it is at all sensible, I prefer to acquire raw data, rather than depend on published compilations or correlations of part of the information. This maximizes the possibility that the data are comparable and that the correlations are meaningful.

2 Measuring the Growth of Literature

Once we begin to study the growth of scientific literature we shall be enmeshed in numbers. Before doing so we should pause to see whether it is worth the effort. Does it make much difference whether one field is in a steady state and another is exploding? Should the student and the parent, the scientist and the administrator be concerned? Indeed so; the general character of the life of a scientist is affected to an astonishing degree by the doubling time of his particular subfield. The intensity of change can vary from the heady pace of the stock market speculator to the stagnation of a bank manager in a depression. We can calculate some specific effects and then surmise how they influence scientific careers. We deal only with averages. Genius may flower among taxonomists, and some molecular biologists are not as smart as others. Moreover, luck affects us all. Consider two specialists of identical merit but born twenty years apart. One is just the right age to replace Professor Smith when he retires from Harvard, and the other is too old. The lucky one gets the job. It may develop, however, that a student's choice of a subfield is one of the most important decisions of his career. If the older man had picked a specialty when it was new and growing rapidly, then he, rather than Professor Smith, might have had the job in the first place.

KNOWING PEOPLE

A scientist knows many people outside his subfield, but we are concerned only with those in it. We assume, until it is shortly demonstrated, that different subfields are growing at highly variable rates, and for illustration we shall consider three models in which doubling times are at the average of 15 years and the extremes of 5 and 45 years.

The professional life of a scientist may be taken as beginning when he becomes a graduate student and ending 45 years later when he retires at 65. Thus, in a subfield in which the literature doubles in 45 years, the number of scientists is constant because new appointments merely balance retirements. Typically, such a subfield might have 200 specialists and it never varies. Nor can it long continue. Either the doubling time increases, or the scientists double their output each 45 years.

In a normal field doubling in 15 years, retirements amount to only one-eighth of new recruitments during a 45-year career and thus are almost negligible. The number of new men increases rapidly. If the initial number is 100, it is 200 in only 15 years, and after 45 years it is 700, allowing for retirement. Such a subfield, starting smaller than a stagnant one, would rapidly outgrow it.

The growth of a subfield doubling in 5 years is truly spectacular. If only 10 men transfer into it at the beginning, there are more than 1,000 in 35 years.

These growth rates have major effects on the age distribution among the specialists in a subfield, and thus on the queuing position of a new man after a given time. The median age in the slow subfield is halfway from beginning to end, or 42 years. In the average and fast subfields it is the initial age plus one doubling period, or 35 years and 25 years, respectively. This means that in the fast field a student is at the median age when he receives his doctorate after 5 years of graduate study. So rapidly does the field expand that he is in the senior one-eighth by the time he is 35, whereas in both the average and slow subfields he would have to wait until he is about 60.

The effects are fairly obvious. The young man in a slow field joins a large population of established, middle-aged professionals and can expect to go through a long apprenticeship. Meanwhile, all the positions and perquisites of academic, professional, and economic power are out of his reach for 20 to 40 years. By the time he receives them he will use them to continue the rule of middle-aged and older men. Professorships and prestigious research appointments are much sought after, because they become available only at long intervals when a retirement occurs.

In the average subfield some of the more energetic and able men may achieve influence while still in their thirties. Thus they introduce youthful views into the development of the subfield, and innovation is more common. Moreover, such young men may be invited into policy-making positions in government and industry and thereby have a much wider influence.

In the fast subfield, positions on national and international advisory committees are held by men in their twenties and thirties. Students are put in charge of oceanographic expeditions and given time on high energy accelerators. New men are vigorously pursued by great universities seeking to establish themselves in the new subfield. Job offers are abundant and rapid promotion virtually assured. A choice of careers is open in research, teaching, administration, and industry and government management of the expanding subfield. Youthful honors are not uncommon, and the recipients have the possibility of highly distinguished careers.

KNOWING THE LITERATURE

Economic well-being and fame outside the profession go to the man in the dynamic subfield. What of his reputation among his peers based on his publications and references to them? Does, perhaps, the productive man in the slow subfield stand just as well in this judgment, which is said to be the most precious one of all? He does not.

We can estimate the literature existing in a subfield as approximately that produced in the three previous doubling periods by the average population of specialists during the period. In our fast model there are 10 people and thus a total literature of 190 papers. In the average subfield there are 100 people and almost 6,000 papers. In the slow model we have to assume that an equilibrium population of 200 has just been established, and in the past the population and literature have expanded at the same rate; the literature consists of 27,000 papers.

We may now consider the activities of a student entering into graduate study in one of these subfields. A diligent speed-reader, he plunges into the literature of the fast subfield and emerges 38 days later, tired but *au courant*. He is prepared to step into the research front without further delay and may be publishing before he gets his Ph.D. In the average subfield a student would have to read 5 papers a day for 3 years to catch up. Much of the literature lies fallow and unread, but even so a significant proportion of the life of a graduate student would be devoted to the library. What then of the student in the old and slow subfield who is confronted with 27,000 papers? They hang over him for his entire professional life, particularly if priority in assigning names is involved. Not only students but also skilled senior investigators are kept busy compiling ever more massive bibliographies.

We can also consider the growth of the literature during the 5 years a student is engaged in graduate studies. Another 190 papers come out in the fast field, and the student absorbs them gradually. Thus he has seen half of the literature develop around him. In the average subfield another 1,200 papers have appeared, or about 1 every second day, which is simply too much to read. The situation is even worse in the slow subfield in which the literature increases by almost 3,000 papers. From the very beginning of their careers, therefore, specialists in average and slow subfields are losing ground relative to the growing volume of unread literature. Their attitudes toward the literature must be quite different from those of a student in a fast subfield.

The situation changes with continuing expansion. Twenty years later all the subfields have a backlog of literature too large to be read by a student, and each is increasing by 3,000 papers in 5 years. Ten years later still, the fast field would acquire 12,000 papers during 5 years of graduate study. Long before, it would have broken into a group of narrower and diverging sub-subfields. It appears that the time for a student or young scientist to get into a fast subfield is somewhere between the third doubling period, when it is identifiable, and the sixth, when it begins to be unwieldy.

The specialist in each of our model subfields has now received his doctorate. How will his career develop? Consider the effort necessary to become established in the subfield. At 3-papers-per-year productivity, and the age distributions we have calculated, a new man in a fast subfield need only write 6 papers before arriving at the median age. In an average subfield 36 papers represent the effort of the middle man in the queue, but in the slow subfield it is 57 papers by age 42. Thus the sustained effort to become established as a sound experienced specialist varies enormously. The disparity in necessary effort increases with time. After only 36 papers, a man

Measuring the Growth of Literature

in a fast subfield is a patriarch in the oldest one-eighth of the population. In the average and slow subfields, 110 papers are required.

Even with single-minded dedication, the production of scientific papers cannot be sustained for a lifetime by many. At some point the scientist drops out of the race to enter a different one or to become a spectator. Few people write 50 papers, and only fifty American geologists have written more than 120 papers. Consequently, most people in average and slow subfields drop below the normal level of scientific productivity before reaching the midpoints of their careers. Hardly any of the senior group are still active in research. In contrast, the senior men in a fast subfield are much younger, and most engage in research even though they are also administrators or committeemen. The administration and evaluation of research in the fast subfield consequently is guided by people engaged in it, which presumably is as effective as the system can be.

Most scientific papers are ignored from the time the ink is dry. Those that are cited give a far better indication of who and what subfields are receiving notice. We shall find in Chapter 5 that frequency of citation is correlated with rate of growth of a subfield. For our model subfields we can visualize the effect by assuming completely random citations. That is, assume all published papers have the same chance of being cited, and thus the probability that a new one will be cited in a given time is a function of the number of papers and of their rate of increase. In our models the slow subfield also has the largest initial number of papers, and each yearly increment has little effect on the total. Ten years after a paper is published it has only an average of 16 chances in 100 of being cited. Under the same conditions a paper in a subgroup with a normal growth rate has 38 chances in 100. In a fast subfield the chances are 79 in 100. Thus a paper in a rapidly doubling subfield has five times as much chance of being cited as in a slowly expanding one.

It appears that the growth rate of a subfield may have a number of effects on the career of a scientist, but do they really add up to anything significant? Consider first the salary structure for earth scientists. It has many distinctive characteristics. In the first place, the median salaries are lower than in any other physical or biological science, and it may not be entirely chance that geology and related fields on average have long grown relatively slowly.[1] Likewise, the highest decile salaries are lower than in all other fields, but those in the lowest decile are near the middle. Presumably this reflects the relatively high fraction of advanced degrees and therefore of starting salaries in the earth sciences. The correlation between median salary and growth rate of subfields is also rather interesting. Lowest by far is geography, above it is a category called "other earth sciences," and in order above are paleontology, geology, geochemistry, hydrology, oceanography, geophysics, and geodesy. To a close approximation, the faster the subfield is growing, the higher the median salaries. This is just the opposite of the correlation that would be expected from the age distributions. The median age is highest in the subfield that grows slowest. Consequently, a much younger group of oceanographers and geophysicists are getting more money than a much older group of geographers and paleontologists. Money goes where there is an excess of demand not of supply.

Growth rates affect salaries, but a dedicated scientist may not think money is important. What of prestige? This can be evaluated in terms of the age at which scientists in different subfields receive the same honors. We frequently notice, for example, that the Nobel laureates in a given year have very different ages. Physicists are generally younger than recipients of the literature prize, and so on. The largest body of suitable data in the sciences consists of the age at election of members of the National Academy of Sciences. If we consider geology alone, it appears that prior to 1930 the significance of the data is doubtful because of the initial effects of formation of the academy and because of changing patterns of rates of growth of subfields. By 1930, however, the older subfields were generally dormant, and the newer were distinctively more vigorous.

Between 1930 and 1949, 22 members were elected in slow subfields and 11 in fast ones.[2] The ratio reversed in the next two decades, in which 22 members were again elected in slow subfields but 40 in fast. The distribution of ages at election during the whole 40 years shows a striking difference in the two groups (Fig. 2.1). Mem-

[1] B. C. Henderson, 1969, "Manpower and salaries," *Geotimes 14*, no. 2, 13–15.

[2] All members of the geophysics section are considered to be in rapidly growing subfields, and in fact they are mainly in geophysics, geochemistry, oceanography, planetary physics, and meteorology, of which all but the last are certainly growing rapidly. In addition, I have made my own evaluation of the principal specialties of the members of the section of geology and classified them into subgroups.

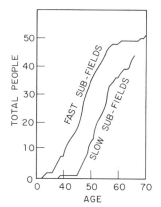

Fig. 2.1 Age of earth scientists at election to the National Academy of Sciences in subfields growing at different rates.

bers in fast subfields are elected about 8 years younger than those in slow ones. The median age for one group is 47 years, and for the other it is 54. Six members in fast subfields were elected at a younger age than anyone in slow ones. Oddly enough, the fast subfields also include the seven oldest members at election. In terms of our model subfields, the members of the fast groups begin to be elected as they enter the senior eighth in their subfield. Their specialties have passed through three doubling periods. In the average subfields the number of doublings is about two, and it is less than one in our very slow model.

We conclude that growth rates in specialties do indeed have a significant correlation with scientific achievement. This may reflect deliberate choice on the part of able young scientists, or it may merely be the effect of exponential growth rates. There must be some significance to the fact that the two youngest members of the group are the only two geologists to serve on the President's Science Advisory Committee. It is distasteful, however, to think that most of us are mere pawns, predestined by a youthful choice of a specialty to a certain type of life. It is particularly heartening to note, therefore, that George Gaylord Simpson, a specialist in the very slowly growing subfield of vertebrate paleontology, was elected at the age of 39. Since that time, he has received countless honors in the form of medals, prizes, and honorary degrees. His achievements show that the menace of the subfields can be overcome.

SATISFACTION IN A CAREER

We have seen the outlines of two career patterns, one prosperous on average, the other less so. Swept along in a dynamic specialty, a scientist has his professional goals within his reach. As a student he receives support from the generous research grants of his professors. He publishes several papers and is an established professional about the time he receives his doctorate. He is showered with job offers from prestigious institutions seeking to expand into his specialty. These continue after he is employed, and his promotions accelerate. He is invited to speak at international congresses and hobnobs with the mighty. We have no way of determining whether he is also happy, but unless he is unlucky, he does not lack for money, position, and prestige beyond that available to most men. Presumably he will see little reason to complain about his career.

The lot of the specialist in a dormant subfield is very different unless he is lucky. His professors have little money for his support. He is fortunate if his thesis is issued as his first publication a few years after he receives his doctorate. Every department already has a man in his specialty, so openings are available only when retirements happen to occur. It is highly unlikely that he will receive many job offers, and he may have to abandon his hopes for research in his specialty in order to take a job teaching geology or science in an undergraduate college. Promotions will be slow or average. If he can do research, perhaps on his own time in summers, he can look forward to long years before he rests easy among the masters of his subject. He may attend the odd regional symposium now and then, but he is not apt to find himself exchanging toasts with members of the Akademiia Nauk or fellows of the Royal Society. We have no way of telling whether he is happy. It is quite possible that he loves teaching, loathes jet planes, and likes a stable life. If he is happy, however, it is despite the lack of money, position, and perhaps even the prestige available to most men. Why should he be content with his career?

It is difficult to see how the life of a scientist in a dormant subfield is very different from the life of any other professional man, except that he is not paid anywhere near as much as most. Poorly paid, little honored, overworked, how do such men react to life? This is a particularly important question at the present moment

Measuring the Growth of Literature 25

because of the prospect that American science in general may stagnate for lack of funds and recruits.

COUNTING PAPERS

Knowing that growth rates are important to scientists, we can plunge into the techniques of measuring them. Defining "science" as the content of scientific journals, we need merely determine the amount of material published at various times to establish a growth curve. The number of journals is itself a reasonable composite measure for the growth of science as a whole. From 1750 to 1950 and perhaps beyond, the cumulative total of scientific journals ever published increased from about 10 to about 100,000 with a steady doubling time of 15 years.[3]

For individual sciences we could simply examine the titles of the journals and separate them into physics, chemistry, and so on. However, this would give only a very rough measure, and it is more satisfactory to count or estimate the number of papers published annually in a given science and thereby achieve more resolution. For many decades only a trivial effort has been required to plot the growth of physics and biology, because the pertinent abstract journals number each entry. It is somewhat more difficult in most other fields, such as chemistry, because the abstracts are not numbered. In zoology, geology, geography, and some others, abstract journals hardly exist, or are only very recent or split into a multitude of uncoordinated subdivisions. The growth of these fields can be obtained from bibliographies, indexes, and acquisition lists, but only with a considerable effort. Prior to the twentieth century, the growth of scientific literature is generally more difficult to analyze, both for lack of comprehensive bibliographies and because many fields of natural science that would now be separated were lumped together. Fortunately, the volume of older literature was relatively small. For most purposes the average growth of science or even of one of the sciences is not enough. Various subcomponents grow at different rates, and the interplay among them must be understood for predictions. This is especially important for rational career selection, because most scientists do all their research in a subfield or

---

[3] Derek J. de Solla Price, 1963, *Little science, big science* (New York, Columbia University Press), p. 8.

even in a narrower specialty. Rapid average growth is little consolation to a specialist in a subject which is in the doldrums. Indeed, it may multiply his frustration. Consequently, a more sensitive measure of growth is needed.

AVERAGE GROWTH AND CITATION AGES

A published scientific paper is merely something written. The only enduring record that it is also something read and used occurs as a citation in a later scientific paper. Citations thus provide a measure of the quality of a publication, and they are much analyzed for many purposes. The easiest technique is to take the citations in one or several issues of a journal and group them according to age. The citations so measured have a common characteristic: averaged over any scientific field and at any time, the number of citations rapidly decreases with age. The decrease is exponential with halving times of roughly 15 years, although it is as little as 5 years in experimental cell research.[4] This observation was commonly interpreted as showing that older literature is relatively uncited and that the lifetime of some of it is amazingly short. However, the method of analysis may be misleading because it takes no account of the undoubted growth of literature. Thus, most of the citations may be to recent literature simply because most literature is recent.[5] Price made a first approximation that the half-life of citations is inversely proportional to the doubling time of literature, and thus the literature of a certain age has a constant chance of being used in later publications regardless of the elapsed time.[6] We may take it that this is demonstrated for the average growth of science and sciences. We could thus use the easily acquired citations to give us the average growth rate if we did not already know it from the more difficult counting of papers. What we shall now examine is whether the age of citations is not a refined measure capable of distinguishing rapid, average, and slow growth and of measuring changes in growth rates of even sub-subfields when they occur.

[4] Paul Weiss, 1960, "Knowledge: a growth process," *Science 131*, 1716–1719.
[5] S. J. Goffard and C. D. Windle, 1960, "Life of scientific publications," *Science 132*, 625.
[6] Price, *Little science*, p. 81.

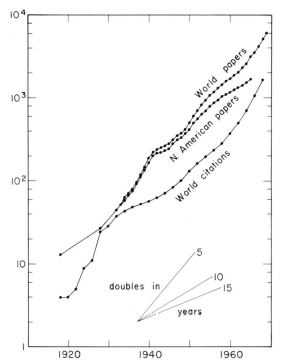

Fig. 2.2 Comparison of indexed papers and citation ages in marine geology.

RAPID GROWTH AND CITATION AGES

Papers in marine geology and geophysics have doubled on average every 5 years since 1920 or ten doubling periods (Fig. 2.2).[7] This is equivalent to 100–150 years of growth by normal science. Except for a retardation during World War II, the growth would be even faster. This very rapid and sustained growth can be compared with the age of citations in papers in marine geology and geophysics.[8]

[7] Annual and cumulative volumes of the *Bibliography of North American Geology* issued as *Bulletins of Geological Survey*; annual volumes of the *Bibliography and Index of Geology Exclusive of North America*, Geological Society of America. These were combined in 1969 into the *Bibliography and Index of Geology* by the Geological Society.

[8] The journals are: (1) January and February 1969, *Journal of Geophysical Research* (*JGR*); (2) January, February, March 1968, *JGR*; (3) January, February, March, April 1967, *JGR*; (4) all of 1960 *JGR*, plus *Bulletin of the Geological Society of America*; (5) all of 1955, selected reprints of articles by eight marine geologists. The decreasing volume of literature necessary for a reasonably sized sample in each time interval is a fair example of the growth of the subfield.

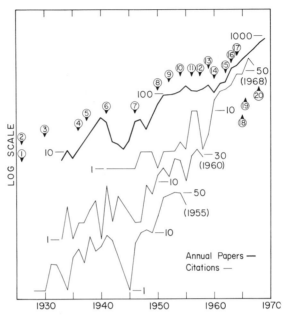

Fig. 2.3 Annual production of indexed papers in marine geology compared with number of citations of a given age in papers published at different periods, and some notable discoveries and events. (1) 1905. Scripps Institution founded; (2) 1925, Meteor Expedition; (3) Woods Hole Oceanographic Inst. founded; (4) Daly's origin of submarine canyons; (5) Kuenen's experiments; (6) Shepard and Emery on canyons; (7) Guyots; (8) Mendocino fracture zone; (9) Grand Banks turbidity currents; (10) Oceanic heat flow; (11) Midocean Rift; (12) IGY; (13) Physiography of North Atlantic; (14) East Pacific Rise; (15) Sea-floor spreading; (16) Vine-Matthews hypothesis; (17) Oceanic tholeiites; (18) Transform faults; (19) Symmetrical magnetic anomalies; (20) Plate tectonics.

The cumulative number of citations older than a given age halves every 5 years before and after World War II, but shows a longer period of slower growth in the middle years. This may merely reflect the small number of citations more than 30 years old even in a sample of 1,630 citations. In any event, the decay in citation ages is strikingly parallel to the growth of papers in this rapidly expanding field. Indeed, the citation ages may be a more accurate measure of growth than the number of indexed papers, because the indexing is full of duplication and subtle changes in categorization. Citation ages do not suffer from these difficulties; the only problem is the one of obtaining a large enough sample of old citations in a small field.

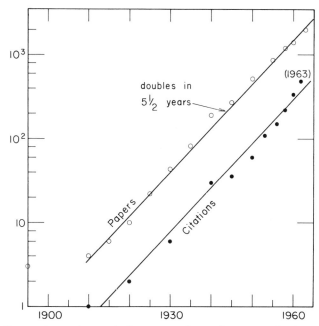

Fig. 2.4 Comparison of cumulative papers in nuclear astrophysics with age of citations in the same field in 1963.

This problem can be solved by comparing the annual output of papers with the citations to each year (Fig. 2.3). The annual output shows a rapid rise to 1940, a drop during the war years, a rapid rise during the late 1940's, a constant output during the 1950's, and a very rapid rise during the 1960's. The citation ages of pertinent papers published in 1968 reflect the recent rapid growth and suggest the stagnation in the fifties, although the number is very small. We can readily magnify and examine these small numbers by plotting the ages of citations in papers published in 1960 in addition to those in 1968 (Fig. 2.3). The curve is quite unlike that of the later papers. Citations to a given year do not decrease rapidly but instead fluctuate around a value of 15–20 for the period of the 1950's, just like the output of indexed papers in marine geology. The citation ages also show the rapid growth in the late 1940's and the minimum during the war with the preceding peak about 1940. However, by that time the number of citations is quite small. Once again we can magnify the growth of the field by studying the citation-age distribution of papers published closer to the period of interest. Papers in 1955 show

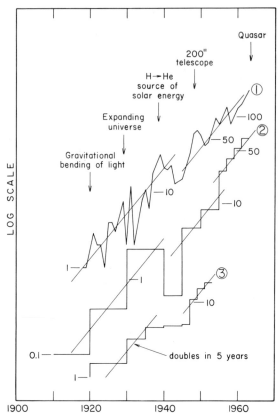

Fig. 2.5 Comparison of annual production of papers in: (1) Nuclear astrophysics with citation ages in papers published at different times; (2) Citations in 1963; (3) Citations in 1953; (4) Some related events.

the rapid growth of the late forties, the minimum growth of the war, and mirror the earlier growth back to 1930. Thus, the citation ages closely reflect the growth history in marine geology and geophysics through numerous complexities.

The field of nuclear astrophysics has also grown very fast for a long time, and once again increase of papers and the decay time for citations are closely related.[9] The field included 4 papers in 1910, when it began a steady doubling about every 5½ years for at least eleven doubling periods before the bibliography was issued (Fig.

[9] Data on papers are from B. Kuchowicz, 1965, *Nuclear astrophysics: a bibliographic survey*, parts 1 and 2 (Warsaw, Nuclear Energy Information Center). Citations are from papers in the *Astrophysical Journal* that are listed in this bibliographic survey.

2.4). Citations in papers on nuclear astrophysics in 1963 have an age distribution with the same slope except for apparent steepening from 1960 to 1963 and an accentuated flattening during World War II.

The annual output of papers in nuclear astrophysics (Fig. 2.5) is quite irregular, but it appears that an exceptionally long if not deep decrease occurred during World War II. Similar trends can be identified in citations in papers in this field in 1963, but the war effect is no more than a pause in citations in papers published in 1953. The sample includes all papers in the field for the given year and published in the *Astrophysical Journal*. A less homogeneous and larger sample including other journals might show an even closer correspondence between papers and citations, but a generally parallel and very rapid expansion is evident. We confirm, therefore, that citation ages give a reliable measure of the growth of science when it is rapid as well as when it is normal.

SLOW GROWTH AND CITATION AGES

Fields of science with slow growth have been relatively rare, but are of particular interest because of the possibility that they may soon become common. We can examine two geological examples. A remarkably long history of vertebrate paleontology publications can be compiled from a series of superb special bibliographies as well as the usual general ones.[10] The first papers identified were published in the middle of the sixteenth century (Fig. 2.6), and there were about 1,000 by 1800. Then the field began a normal doubling in 15 years, which continued to about 8,000 papers in 1850. A decline followed, and by 1870 the field locked into an exponential growth, doubling in 35 years, which continued to 1927, when there were about 50,000 papers in vertebrate paleontology exclusive of North America. Broadly speaking, the field grew normally only when evolution was the center of controversy, and it has been relatively quiescent for the remainder of the last four centuries. Even so, there are more than 50,000 papers.

[10] A. S. Romer et al., 1962, *Bibliography of fossil vertebrates exclusive of North America, 1509–1927*, Geological Society of America Memoir 87, and subsequent volumes dealing with later publications. Memoir 87 alone is in two volumes. I have considered only the publication date of papers by authors with last names beginning with "A." The pages involved indicate the sample is one-nineteenth of the whole, and it has been multiplied accordingly.

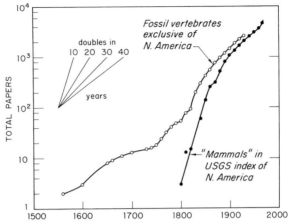

Fig. 2.6 Growth of papers in vertebrate paleontology for four centuries. The growth rate was normal only for a brief period when evolution was discovered.

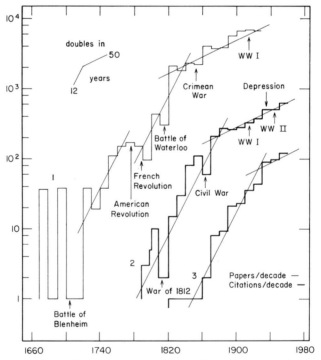

Fig. 2.7 Comparison of papers in vertebrate paleontology and the age distribution of citations in the same fields: (1) Vertebrates exclusive of North America; (2) Mammals in North America; (3) Citations, 1960–1961.

Papers indexed as "Mammals" in the bibliographies of North American geology show a normal rate of growth continuing to about 1900, which probably reflects the great excitement of the discoveries in the matchless exposures of the western deserts. Nevertheless, in 1900 this field also became quiescent in America, with a doubling rate of 35 years, like that of international vertebrate paleontology. This is still continuing.

The outputs of papers per decade in vertebrate and mammal paleontology also can be resolved into two trends, an older one with a doubling time of 12 years and a more recent one doubling in 50 years, although the changes in trend occur at different times (Fig. 2.7). The worldwide output of vertebrate paleontologists was sensitive to major wars such as the Napoleonic wars, and, in the United States, the War of 1812 and the Civil War. These all occurred during the period of relatively rapid growth and thus are analogous to the effect of World War II on fields that were expanding rapidly at that time. On the other hand, the world was not with vertebrate paleontologists either late or soon during the long period of very slow growth. The average output per decade was unaffected by either of the world wars or by the great global depression between them. A simple explanation for this is that scientists, or anyone else, in a slowly advancing field have a relatively old average age. If most people are too old to go to war, their work is little affected by it. Likewise, the low cost of vertebrate paleontology research makes it relatively immune to depression, provided a scientist has a job as a professor as well.

The age distribution of citations in vertebrate paleontology papers of 1960–1961 can also be analyzed as following two trends parallel to the growth of papers in the field.[11] However, the change in trend occurred only 30 years ago and thus does not correspond to the time of a change in rate of output. A single trend of intermediate slope would be an equally valid interpretation. The end result of the differences in trend, or time of trend change, is that the proportion of citations to papers increases markedly with time. The ratio of papers on mammals to citations in vertebrate paleontology is constant at about 6/1 for the period 1930–1960, but in earlier periods

---

[11] Citations are from 21 papers in vertebrate paleontology in the *Journal of Paleontology* in 1960 and 1961.

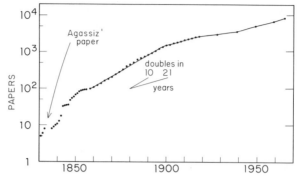

Fig. 2.8 Growth of papers in glacial geology. Growth was rapid after Agassiz published and normal as late as 1900, when it slowed.

it was much higher, about 100/1 in 1860. In terms of growth, the ratio is constant for only one doubling period and then increases rapidly. This is a very marked contrast to the constant ratio in marine geology and nuclear astrophysics, which has persisted for 45–50 years or ten to eleven doubling periods.

We shall now examine one other field with slow growth to see if it has characteristics similar to vertebrate paleontology. Glacial geology provides an example (Fig. 2.8).[12] A dynamic growth began in 1837 when the young Louis Agassiz published his celebrated article on the former existence of an ice age in western Europe. The number of papers then doubled every few years as the evidence for a widespread glacial age was discovered. By 1858, the growth became exponential with a doubling time of 10 years, and this continued until 1900 while the moraines of repeated glaciations were mapped in North America. Then a very slow growth trend began, which has persisted to the present. The average doubling time has been 27 years with slightly faster growth at the beginning and end of the period.

The annual output shows more details of this growth (Fig. 2.9). For a brief period after Agassiz' paper, it appears that geologists

[12] The output of papers is taken as those indexed in the various issues of the *Bibliography of North American Geology* under the categories of "glacial————." Citations are from papers in the same field in the *Bulletin of the Geological Society of America* plus the *Journal of Geology*, 1960–1961; and in the *Bulletin of the Geological Society of America*, 1890–1893.

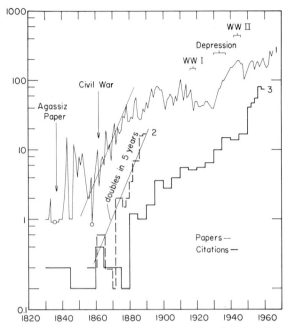

Fig. 2.9 (1) Annual output of papers in glacial geology and geomorphology; compared with citation ages in the same fields in the *Bulletin of the Geological Society of America* and *Journal of Geology* in (2) 1890–1893, and (3) 1960–1961.

were reevaluating the meaning of their observations. Then the number of papers increased fifteenfold momentarily before a 20-year decline began. This was followed by a 25-year exponential growth with a doubling time of only 5 years for the annual output. Next the output was relatively constant, that is, the growth was arithmetic rather than exponential, until the 1930's, when a step increase occurred, and then another two-decade period of constant output began. Perhaps the output is again increasing at present.

This record, like that in vertebrate paleontology, seems to pay little regard to current events. The fluctuations during the world wars appear little more than random, and during the great depression the output of papers tripled instead of dropping. Even during the period of rapid growth, the Civil War had no exceptional effect.

If we compare output with citations we also see an effect resembling that for vertebrate paleontology. The ratio of papers to citations in journals published in 1960–1961 is 20 in 1900, near the beginning of the steady state or arithmetical growth; it decreases to 6 in 1930 and 3 in 1960.

The citations in papers published in 1890–1893 have an entirely different relationship to the number of papers published. At that time, the field was still in a phase of rapid exponential growth, and the ratio of papers to citations behaved as it has since done for marine geology and nuclear astrophysics. The age distribution for citations has the same slope as the growth curve for papers published. The ratio of papers remains relatively constant at about 8–10, while the annual output increases fivefold. Thus, in a single field we find a confirmation that the distribution of citations is a reliable measure of rapid growth but is less so for steady-state or very slow exponential growth. It has been observed that in such fields as paleontology and zoology there is a marked tendency to cite very old literature compared to physics or chemistry. This occurs only because such a large fraction of the literature is old. Scientists in normal and rapidly expanding fields sample all the literature in the field randomly in time. The proportion of papers cited to those published in a given year does not vary. On the other hand, scientists in slowly growing fields escape the tyranny of a vast backlog of literature by ignoring most of it. Perhaps the output in a dormant field is pretty dull stuff.

ANATOMY OF AVERAGE CITATION CURVES

We have arrived at the empirical conclusion that the distribution of citation ages in a journal is related to the growth of a field, but we have made no attempt to consider the origin and structure of the distribution. A journal is a collection of scientific papers, and the citations occur within the papers. The distribution of citation ages in a journal is merely a composite of those in the papers, which themselves may be quite different. For example, the *Astronomical Journal* for 1968 contains citations which in composite indicate an age distribution with an increasingly steep slope from 1900 on; so does the *Bulletin of the Geological Society of America* for 1970. Thus, there are many citations to recent work but few to the less voluminous work of the past. This might seem to suggest that all the papers do likewise, and scientists are universally concerned more

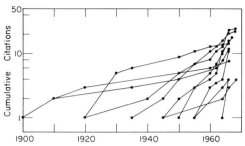

Fig. 2.10 Age of citations in individual selected papers in the *Astronomical Journal*, 1968.

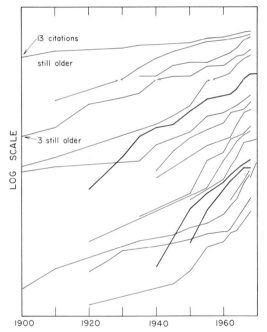

Fig. 2.11 Age of citations in individual papers in *Bulletin of the Geological Society of America,* Jan. 1970, vol. 81 (stacked to display slope).

with the present than the past. The age distributions for individual papers, however, show something quite different (Figs. 2.10, 2.11). Some astronomers and geologists cite only literature of the last four years, but one astronomer cites five times more references in the 1930's than the 1960's. Another cites none in the 1960's, eight in the 1950's, none in the 1940's, and five in the 1920's and 1930's.

One geologist cites two papers each in the 1960's, 1950's, 1940's, and 1930's; another cites six papers in the 1940's and 1950's, and the same number in the 1840's and 1850's with less frequent citations during the intervening 90 years. Whether they have few or many citations seems to make little difference to the slope of the age distribution in a given paper. Review papers tend to have an even distribution, but half the citations in one are less than 3 years old and in another the half-life is 30 years. Some papers cite more current literature and some cite more in the past.

What is a citation? No more than a reference to related work in the past. In a very slowly growing subject it may happen that no closely related publications have appeared for several decades. Why would a scientist who takes up the subject have occasion to cite anything written during the hiatus? He might cite a broad synthesis, but if so, it would have some part related to the subject. Some of the observed distributions suggest that a subject flowered for a decade or two or had a flurry of excitement for only a few years. The citations in papers in that subject and at that time would have had very short half-lives, because all the papers cited had just been written. After a long quiescent period, the half-lives of citations in the subject would be very long, because they would include the same but now aging papers.

All this leads us to the general conclusion that scientists are conscientious and diligent scholars who cite earlier literature fairly and without regard to its age. The sole exception we have found is in fields with very slow growth in which the older literature is neglected. Perhaps this is not merely because the subject is dull. It may be that no one has ever worked on a related subject in the century since some of the older research was done. Presumably most of the insects and fossils described in the nineteenth century simply are not likely to be of interest to anyone. Many a vertebrate paleontologist could describe a dozen new species a year, but if no other examples were ever found his colleagues would little note nor his successors long remember. Thus, in a field in which papers are cited infrequently, many may never be cited at all.

In a whole science at any given time most of the fields and subfields are normal or growing rapidly, even though some may become dormant in the future. Consequently, at any given time the citation-age distribution in a whole major journal will show that most are relatively recent. This is because most of the related work is recent, not because of a disproportionate citing of recent literature.[13]

[13] Derek J. de Solla Price presents a different analysis of what he calls the "immediacy factor" in *Science 149*, 510–515 (1965). He deduces that 70 percent of the citations are randomly distributed to all the scientific papers ever published. The remaining 30 percent are "highly selective references to recent literature." His study is based on the broadest possible composite of papers and citations compiled in the *Science Citation Index* data for 1961. His observations are confirmed in many of the studies in this book, but they do not apply to all subjects at all times. The undoubtedly real "immediacy effect" applies to the subject matter of the papers rather than to the citations within them.

3 Growth of Sciences

We may accept the fact that science as a whole and individual sciences as well grow exponentially, so that the literature doubles every 10 to 15 years. Nevertheless, the growth is not unvaried, and a study of the spurts and pauses reveals much about the factors that influence it.

Variations in growth are best seen in the annual rather than cumulative output of scientific papers (Fig. 3.1). During the last 70 years biology, chemistry, geology, and physics have grown with vaguely parallel trends.[1] A period of relatively slow growth occurred from 1900 to the 1940's, and much faster growth followed to the present. The variations in timing and rate of growth are revealing.

From 1900 to 1920, geology and physics had an average steady-state output that is equivalent to arithmetical rather than exponential growth. This was climaxed by a 25-30 percent dip during World War I, which had an obvious cause. However, the steady-state condition started 10-15 years earlier, and it presumably had some other cause. The growth of chemistry is in marked contrast. Abstract data began in 1907, when output was about 9,000 papers annually. In 1920, it was 21,000 annually or more than double. The effect of

---

[1] *Biological Abstracts* and *Physics Abstracts* number abstracts and thus are easy to compile. The five- or six-digit numbers may induce statistical euphoria, however, because of duplication in such classifications as biological physics. These errors are rather small. In *Chemical Abstracts* they are larger, because abstracts are not numbered until recently, although columns are. Abstracts per column have been sampled at frequent intervals to estimate annual output. Despite little variation in page size, the number per page steadily decreases because of cross-referencing to multiple authors who are steadily increasing. In geology, in which I include all the earth sciences, annual output of papers is estimated by sampling the number per page in the various issues of the *Bibliography of North American Geology*.

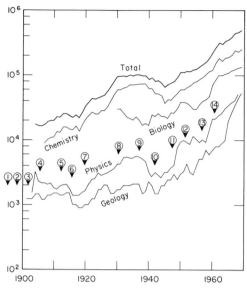

Fig. 3.1 Annual production of abstracted or indexed papers in various sciences. The numbers above the physics papers refer to the following discoveries and inventions: (1) X-rays; (2) Radioactive decay; (3) Planck's quanta; (4) Special theory of relativity; (5) Superconductivity; (6) Bohr atom; (7) Mass spectrograph; (8) Cyclotron; (9) Isotope tracers; (10) Atomic pile; (11) Digital computers; (12) Hydrogen bomb; (13) Lunik; (14) Laser. Developments in physics from Walter Sheperd, 1965, *Outline history of science* (New York, Philosophical Library, Inc.).

the war was merely to stall the growth from 1914 to 1918 without any significant dip—this despite the war commitments of chemists and the chemical industry in almost every important industrialized nation. The flower of English and French and German science went into the mud of Flanders, but 15,000 papers a year still poured out.

In the 1930's all the sciences grew rapidly although at different rates and with a brief pause in physics and chemistry in mid-decade. The next two decades saw another period of general average stagnation in that the annual output was about the same in 1950 as in 1930. Once again the timing and extent of this pause in exponential growth varied with the different sciences. Geology output went steady-state in 1926 and stayed that way to 1945. A dip occurred briefly during World War II, but it was hardly more pronounced than in 1928 and 1938.

Biology abstracts begin in 1930, and throughout the decade they steadily drop. During the war they first rose and then steadied at an average of 24,000 papers per year from 1942 through 1945. The war caused only a pause in growth not a dip.

In chemistry the growth of the twenties ceased about 1930 and output remained steady-state for a decade. Through the war years it slowly sagged, until in 1945 it was down by 50 percent. Physics followed a similar pattern, but began to sag a little earlier and dipped more abruptly, but by almost exactly the same amount.

From 1945 to the present all the sciences have grown enormously, and the smaller they were originally, the faster it has happened. Chemistry and biology have increased sixfold to an annual output of 250,000 and 135,000 papers respectively. Physics has grown fifteenfold to 50,000 papers annually[2] and geology twenty-sixfold to 53,000, but starting from a smaller base in 1945 than physics. The same relationship is observable if present output is compared to the peak in the 1930's. The increases are: chemistry, fourfold; biology, fourfold; physics, tenfold; and geology, which had no peak in the thirties, still increased twenty-sixfold.

These growth patterns for the last 70 years are better known than earlier ones, but they seem similar to the patterns for the last 150 years in glacial geology and vertebrate paleontology. What of a whole science rather than just some of its components?

United States geology has been studied by the herculean measure of counting pages, which provides relatively accurate information for comparison with the population and productivity of scientists. The geological literature of the United States consists of three important components: normal publications in scientific journals, those of state geological surveys, and those of the U.S. Geological Survey.

We first tabulate the number of pages produced in all significant national journals in geology at 5-year intervals and interpolate to approximate the cumulative growth of this literature.[3] Inclusion of all possible literature would not affect the result by much as the number of pages accumulated — from 445 in 1818, the first year

---

[2] A pause in average growth in the 1950's is explained in the section on growth of subfields.

[3] We exclude local journals, trade journals for rock hounds, and the *Pan-American Geologist* because it was mostly full of nonsense.

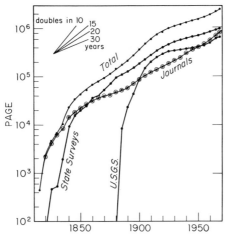

Fig. 3.2 Pages of various components of American geological literature.

of the *American Journal of Science,* to 830,000 in 1968.[4] Growth was rapid until about 1830 and then locked into a simple exponential expansion with a 25-year doubling time until 1955; it then accelerated to normal with a doubling time of about 16 years. From a consideration of only the ordinary scientific journals it seems as if American geology never had a period of normal scientific growth until the middle of the twentieth century (Fig. 3.2).

However, the other components of growth yield a different story. The history of the state geological surveys is remarkably well documented by Merrill and contains a wealth of data on the relationships among science, society, and government.[5] Some geology is included in the state papers of New York as early as 1820, and the first formal state survey report came from North Carolina in 1824. By 1835, when the Geological Survey of Great Britain was founded, nine state surveys were active and 2,123 pages of geology had been

---

[4] Most earth science journals in English contain numerous papers by U.S. scientists regardless of the country in which they are published. For this reason a number of the more international earth science journals were included after 1950 when they began to be common. Certainly the rapidly increasing numbers of such journals published commercially in the Netherlands are almost wholly the creatures of United States science.

[5] G. P. Merrill, 1920, "Contributions to a history of American state geological and natural history surveys," *Smithsonian Institution, Bulletin 109*; 1924, *The first one hundred years of American geology* (New Haven, Yale University Press; facsimile reprint, 1964, New York, Hafner Publishing Co.). Based on an elaborate, official poll of the individual state surveys.

published. By 1879, when the United States Geological Survey began, some 93,000 pages of geology had been published by the states. The work was of such interest that Pennsylvania issued some editions in German for the immigrant farmers.[6]

This state literature increased rapidly for the first 25 years and then settled into exponential growth with normal doubling in 15 years from 1845 to roughly 1910. Then some stagnation began, and the doubling time increased to 40 years and appears to be continuing at that rate. It is interesting to note that the total volume of geological publication by the states surpassed that in normal scientific journals in 1856, and it was more than twice as great 50 years later. If present trends continue, however, scientific journals will once again be dominant by about 1974.

The publications of the United States Geological Survey form the third major component in this literature, and they are a joy to study. Everything is catalogued, indexed, and widely available. These splendid records tell a sad tale (see Fig. 3.2). The number of pages grew rapidly as the classical accounts of the geology of the west were written by a small band of scientific explorers. The growth began to slacken by 1900, but it did not drop below a normal doubling time of 15 years until 1925. From then until about 1955 the doubling time was an appalling 70 years, or considerably slower than normal expansion of either the national population or the gross national product. About 1955 an obvious rejuvenation began, and the doubling time for survey literature dropped to a spritely 22 years. There is a dance in the old dame yet, although hardly a tarantella.

We may now examine the sum of these three components. Pages of geological publications in the United States increased rapidly from 1815 to 1845 and then locked into exponential growth and

---

[6] The page outputs vary so much in the early years that it was necessary to count them for each year until 1910. For example, the state survey of Arkansas wrote no geology in 1890, 2,608 pages in 1891, and 1,164 pages in 1892. Pages were compiled only at 5-year intervals after 1910, because the output became less variable. Almost every state has had a geological survey from time to time, and publications have appeared in such places as the state legislative record and are difficult to acquire. We have solicited each of the state surveys for a list of publications with page numbers, and it appears that some of them are not available. We have also examined the geological literature in two university libraries. Even so, some literature has escaped us, but we estimate that it is only a fraction of a percent.

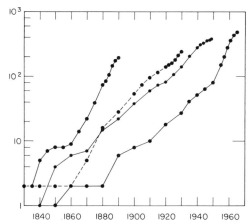

Fig. 3.3 Cumulative plot of citation age distributions in different years of the *Bulletin of the Geological Society of America*.

doubled every 20 years until about 1880. Then growth was still exponential but at a normal pace until about 1915. This reflected normal growth of the state literature, slow growth of the journals, and rapid growth of the Geological Survey publications. From 1925 to about 1955 the doubling time increased to 38 years, and the contribution of the components was reversed. The survey grew hardly at all, the state surveys slowed, and the journal contribution became the one growing most rapidly, although its growth was unchanged. By 1955 or so the upturn in journals and the Geological Survey publications caused an increase in the growth of all literature in geology that still continues. In sum, these data indicate that United States geology has alternated between slow and normal growth at 30- to 40-year intervals. The increased detail available for this century (see Fig. 3.1) shows that the last slow period was very slow, and the present fast period is comparatively fast as though to catch up. If the present trend continues for 30 years more, the growth will average as normal for 150 years.

As might be expected, the age distribution of citations in the *Bulletin of the Geological Society of America* reflects this history of growth (Fig. 3.3). The cumulative slope in 1890 is steep, indicating rapid growth. In 1930 and 1950 it slopes gently for recent citations and then steepens for older ones. In 1968 the slope is again steep for recent citations.

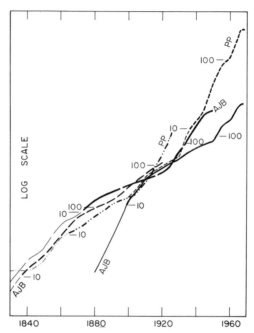

Fig. 3.4 Agglomeration of citations in different volumes of *Plant Physiology* (1930, 1968) and *American Journal of Botany* (1920, 1950, 1968).

We can use the correlation between growth and citation ages to examine botany, another relatively persistent and stable science like geology.[7] Citations from two journals suggest a doubling of botany literature at a normal rate from 1830 or earlier to about 1860, when it slowed to a doubling in 22–24 years. The small sample suggests an annual output that was relatively constant from 1860 to at least 1920, indicating a steady-state or arithmetic growth instead of exponential (Fig. 3.4). In 1920 the *American Journal of Botany* included hardly any components of rapidly growing current research; all was at a sedate pace. In 1930 *Plant Physiology* was dominated by rapidly expanding research, but a major component of the slower research remained. It had almost vanished in *Plant Physiology* in 1968, when 90 percent of the citations were less than 20 years old and the doubling time was only 7 years. In the *American Journal of Botany*, however, the slower components remained important in

[7] Citations have been compiled from the *American Journal of Botany* in 1920, 1950, and 1968, and *Plant Physiology* for 1930 and 1968.

1950 and were still more so in 1968. It appears that rapid growth began in *Plant Physiology* about 1930, just when it was stalling in most fields. In botany the slower component is still important, and arithmetic growth may have continued even as late as 1950, when botany probably joined the general expansion of science. It is interesting to note that whereas the older papers were apparently proportionally cited as late as 1950, they were not cited in 1968. This may be the same rejection of a mass of older literature that we observed in vertebrate paleontology and glacial geology, which were also long in a steady state.

These studies of growth reveal that individual sciences may have long periods of slow and rapid growth. Not just geology: the number of abstracts in physics was the same in 1904 as in 1944, and between those dates it was lower for about the same time that it was higher. The pause in growth in the 1930's and 1940's might be attributed to the depression first, and then the war. Why did geology go steady-state five years before the depression? Why did physics pause during the decade before World War I and the decade of the 1950's? The growth of science is influenced by many factors, and we shall return to these after further consideration of the growth of literature and the population of scientists.

GROWTH OF TRANSITIONAL FIELDS

If we visualize an expanding research front, it should consist of lobes or pseudopoda separated by more or less narrow and deep recesses. The lobes are the principal sciences, and the recesses are the boundaries between them. Thus, zoology advanced in the nineteenth century and physics as well. At that time, zoology was the larger lobe, and it probably was expanding at the same rate as physics. Since then the physics lobe has moved far ahead. At the boundary between them was a recess in biophysics, created because it depended on developments in the main sciences. Even though one lobe outstrips the other, the recess may exist. In a sense this is self-fulfilling. If, for example, someone in biochemistry had discovered an element, it would have been considered an advance in chemistry rather than biochemistry. The only question, then, is about the shape of the recesses; it is not doubtful whether they exist.

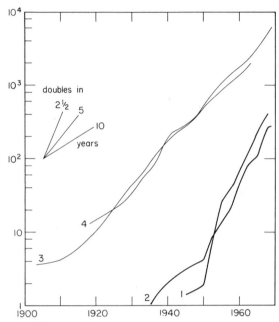

Fig. 3.5 Cumulative totals of papers and citations in several lively transitional fields: (1) Plant physiology, citations; (2) Molecular biology, citations; (3) Nuclear astrophysics, papers; (4) Marine geology-geographics, papers.

Most recesses are quite narrow, although of variable depth. Well-trained, active research scientists are at least expected to be versatile and able to bridge from one field to another when required. A chemist may spend twenty years making analyses and behaving strictly as a chemist — but, not just with test tubes and Bunsen burner anymore. Automatic sampling and analysis with computer control of the modern type would look like mathematical-physical-engineering-chemistry to a strictly test-tube man. Likewise, zoology and mathematics have been separated by a deep recess with some population statistics and d'Arcy Thompson's *On growth and form* at the bottom of it. But the computerized analysis of microorganism species as shapes identifiable by automatic scanning devices is underway.

Because the recesses are so narrow, very rapid advances may be expected when they become deep and then are bridged. A persistent and important group of problems in a science may become tractable because of a discovery or merely a newly appreciated application of

Growth of Sciences 49

work in another science. All of a sudden it is possible to use existing samples or data for analysis by a more revealing technique; or the puzzles of 50 years are unlocked by a new paradigm developed in another field. Many people enter the new transitional field, followed by students trained in it, and it expands rapidly. This has happened during the last 30 or 40 years in molecular biology, marine geology, nuclear astrophysics, and plant physiology, among others (Fig. 3.5).

The new transitional field expands, but what then happens to the preexisting sciences and the boundary between them? If the two lobes of the research front are expanding equally, the new transitional field may fill in the recess, outstrip the marginal subfields in the main sciences, and create a new lobe with recesses between it and the fields from which it was derived. The end result is three healthy expanding sciences and narrow recesses amenable to bridging at any time.

The development is apt to be different if the two original sciences have been expanding at quite different rates. We can visualize one lobe far in front of the other. Slow growth may occur for many reasons, but it implies a detachment or lack of communication with the other sciences in the research front. If a bridge occurs between a retarded lobe "R" and an advanced lobe "A," only a few people in R will be able to understand the new transitional subfield RA, because it uses unfamiliar equations or technology. On the other hand, most scientists in A will find RA straightforward but messy, because it depends on unfamiliar and uncontrolled or structureless data.

If R is only moderately retarded, it may gradually be able to absorb the new students in RA, and with the infusion of new life, the whole science may begin to expand. If it is highly retarded, the members of R may simply disown RA as being a thinly disguised subfield of A. At that point, R begins to become more and more withdrawn. The field is ever more narrowly defined and every startling advancement is excluded as non-R. Things may look much the same; specialists in R continue to win gold medals and are elected to learned societies. Nevertheless, the other learned members are apt to ignore R's because they have no way to communicate with them. We have found that many sciences have grown slowly for long periods, and some of them have become retarded in the meaning used here. Geology until recently was an example. We shall study some of the things that happened to it and other retarded sciences as we progress through the pages that follow.

GROWTH OF SUBFIELDS

In a general way, we can say that "physics" or "geology" grows rapidly at some times and slowly at others. Some understanding of how this may occur can be gained by studying the subfields which make up these sciences. We begin with physics, which is relatively well known, and proceed to geology, which is studied more comprehensively as the basis for detailed comparisons of growth. *Physics Abstracts* is divided into subfields that are somewhat mutable from 1900 to 1970 but can be used to trace growth. The slowly growing components selected are "acoustics" and "optics." The rapidly expanding ones are "atomic-nuclear-elementary particle" and "solid state." The name changes in the abstracts tell an interesting and typical story.[8] In 1900 the slow categories are called "sound" and "light," and the fast ones are in "general physics." By 1921 a subdivision of "light" appears called "radio-activity." In 1930 the subdivision becomes a separate one, and the hyphen is dropped as enough people become interested in the subject to warrant a simplification of the name. By 1940 "sound" becomes "acoustics," but the category is still in alphabetical order after "radiation," which is the new name for "light." "Radioactivity" remains, but a related category of "atomic and molecular structure" appears and is included with it in the graphs (Fig. 3.6). By 1950 the slowly growing fields metamorphose to "vibrations, acoustics," and "optics, radiation, spectra." Meanwhile, "radioactivity" remains, but it now has four related categories. "Theory of solid state" appears as an indexed category, although no papers are published in some months. This is an unusual practice and a brilliant forecast of the future. Categories related to solid state exist but are not yet directly linked in the indexing.[9] The categories are about the same in 1960 except for a further proliferation of subfields in nuclear and particle physics.

---

[8] The abstracts were examined at roughly decade intervals, and all other numbers are by interpolation from graphs.

[9] Those included here are "structure of solids," "elasticity," "strength," "rheology," and "crystallography."

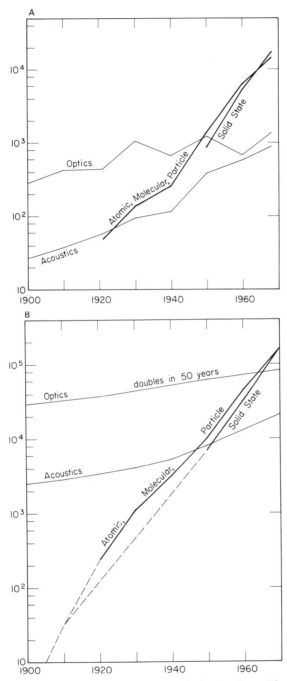

Fig. 3.6 Output of papers in some subfields of physics, from *Physics Abstracts*. A, Annual. B, Cumulative.

By 1968 solid state physics suddenly has about as many subfields as nuclear physics, and it is optics that is becoming confusing. New subfields of quantum optics and ion and particle optics are identified, and they are not lumped with more conventional optics.

According to these categories, the literature in nuclear and related physics has doubled every four or five years since 1920, and solid state physics has done so since at least 1950.[10] In contrast, acoustics and optics have grown much slower and with varying rates. Acoustics had a doubling time of almost 40 years prior to World War II, a rate quite comparable with glacial geology and vertebrate paleontology but not steady-state, inasmuch as the annual output steadily increased rather than remaining constant. A few acousticians grew affluent advising Hollywood about talking pictures, but not too much was going on. During the war, however, acute problems of detecting submarines by sonar caused a great spurt, and the annual output tripled in a decade. The whole open literature began to double at normal rates. That is by no means the whole story, however, because much of the literature is "Classified" by the Navy and does not appear in *Physics Abstracts*. Specialists in underwater acoustics cannot work without access to this classified matter, and they estimate that it is roughly the same size as the open literature.

It is dangerous to consider anything but the literature of normal scientific journals or printed and freely available government publications in evaluating the growth of science. Many industrial chemists or mining geologists write reports of the type that might appear in open scientific literature if they were under different auspices. The work does not qualify as science if it cannot be read widely and used as the basis for further research. Acoustics for submarine detection and physics for warhead development are in a different category, in that an extensive classified literature exists and it is used as the basis for further classified research. This represents an uncertainty in evaluating growth, especially because some of the literature may be declassified at any time and thereby suggest a spurious spurt in output. We should also take this into account when we attempt to evaluate the productivity of scientists.

[10] The production of papers in these categories early in the century was estimated by assuming only one paper of each existed in 1900, which is a reasonable extrapolation of the later trends. For the slowly growing subfields, assume that the number of papers existing in 1900 was 100 times the annual output at that time, which is not unreasonable judging by twentieth-century growth and the antiquity of the subjects.

Growth of Sciences 53

Optics was the subject in which Isaac Newton took the most pride in his achievements, and it has interested many others in the following three centuries. Consequently, the literature is large, and presumably much of the cream has been skimmed, because it has doubled during this century at an extremely slow rate. Indeed, since 1930, the annual output has not grown on average and thus the subfield is in a steady state, occupying the interest of about 15 percent of all physicists.[11] Meanwhile, related subjects have developed as a bridge with the great advances in very different subfields of physics, such as quantum theory. These might have been incorporated with classical optics by the indexers with a resulting spurt in literature output. It has not occurred, however. Optics has been a retarded sublobe in the advancing research front of physics. Judging by the indexing procedures, optics may be in process of losing all its lively aspects and thereby becoming more isolated and retarded in the manner previously discussed.

Considering only these four subfields, we can construct a simple history of physics in this century. In 1900 the literature and the investigators in slowly expanding acoustics and optics were dominant, and the whole science was essentially steady-state despite the developments that have since become so famous (Fig. 3.1). By 1920 atomic and molecular physics papers began to appear in numbers equal to acoustics but still only one-tenth of those in optics. The rate of growth began to increase but only slowly. By the end of World War II, the annual output of the rapidly advancing subfields was at last equal to that in the slow ones; then began the still-continuing rapid expansion of physics as a whole. Even so, the effect of the different rates of growth of subfields may still be felt. In the familiar parlance of Wall Street, in the 1950's there was a mixed market. Some subfields went up and some down (Fig. 3.6). It appears that the long pause in the growth of physics in the 1950's occurred because optics, and perhaps other subfields, had a drop in output rather than that all the subfields paused uniformly.

We turn now to geology. The growth trends of its many subfields can be derived by compiling the number of entries under pertinent indexed categories in the various years of the *Bibliography of North American Geology*. Thus, the growth of local or regional geology can

---

[11] National Academy of Sciences, 1966, *Physics: survey and outlook* (Washington, National Academy of Sciences), p. 78.

be sampled by determining the number of papers indexed under a selected group of states, such as Colorado. Likewise, the growth of invertebrate paleontology can be sampled by considering the growth of papers about certain families or orders. This does not give the total population of a whole field or subfield of geology. Moreover, if we attempt to see how the sample compares with the whole field, it rapidly becomes apparent that the indexes contain many duplications that can be eliminated only with difficulty. Consequently, the analysis of growth of subfields by study of bibliographies depends on a careful selection of index categories, some of which have changed little during the last century and thus automatically apply to slowly changing subjects.

In addition, it is useful to have some indication of the proportions of the subfields relative to the whole of the earth sciences. The first 1,423 papers in the various issues of *Bibliography of North American Geology* from 1785 to 1960 were categorized into a system of subfields and subdivisions.[12] The largest categories are as follows: economic

---

[12] The fields selected and the index categories sampled are listed below. The numbers correspond to those in the pertinent illustrations. Numbers in parentheses are the fraction of the earth sciences represented by the subfield in a separate analysis of 1,423 in the *Bibliography*.

Slowly growing and cyclical subfields:

1. Economic, nonpetroleum (11.2%): building materials, coal, copper
2. Origin of ore deposits (1.2%): origin of ore deposits
3. Economic, petroleum (4.0%): petroleum, natural gas
4. Geomorphology (7.7%): glacial, geomorphology, or physiography
5. Igneous rocks (3.2%): igneous rocks, volcanic rocks, petrology
6. Metamorphic rocks (1.3%): metamorphism, metasomatism
7. Invertebrate paleontology (5.0%): cephalopoda, crinoidia, gastropoda, trilobata
8. Vertebrate paleontology (4.2%): mammalia, man, reptilia
10. Regional and local geology (11.0%): California, Colorado, Connecticut
11. Sedimentation (1.5%): sedimentation
12. Stratigraphy (8.8%): Cambrian, Cretaceous, Tertiary
13. Structural geology (1.1%): faults, folds

Rapidly growing subfields:

9. Micropaleontology (1.8%): conodonts, foraminifera
14. Astrogeology (0.4%): meteorites, moon
15. Continental drift (0.0%): continental drift
16. Evolution (0.1%): evolution
17. Experimental investigations (3.4%): experimental investigations
18. Facies (0.4%): facies
19. Geochemistry (5.6%): geochemistry, isotopes
20. Geophysics (5.4%): geophysics, seismology, gravity
21. Marine geology (0.2%): continental shelf, continental slope, ocean, submarine
22. Paleoecology (0.6%): paleoecology

geology, 16.4 percent; paleontology, 11.7 percent; regional geology, which includes quadrangle mapping, 11.0 percent and stratigraphy, 8.8 percent. Fully 10 percent of this sample are not reports of research but of administrative affairs.

The most widely known fluctuating phenomenon is the stock market, and we can take advantage of some of its familiar terminology. We identify growth, stable, and cyclical fields. Stable fields resemble the American Telephone Company in that they are old, large, and tend to grow constantly but relatively slowly. The growth rates generally are somewhat slower than normal for science and doubling times of 20 years are common. Cyclical fields are more like U.S. Steel in that supply and demand or support or interest appear to fluctuate. Such fields are petroleum geology, astrogeology, and structural geology (Fig. 3.7). Some of these fields have grown large and had long periods of very slow growth. Stable and cyclical fields constitute 79 percent of the sample of the last 150 years of American geology.

Growth fields are quite different in that they generally have persistent growth trends with doubling times of only 5 to 10 years. Such are geophysics, geochemistry, marine geology, paleoecology, and experimental geology (Fig. 3.8). Despite the fact that they are now as large as the older stable fields, they amount to only 21 percent of the whole, because in 1920 they were all almost negligible. This would account for the fact that the number of articles in geology as a whole increased so slowly in the first half of the century. During that period, the mass and slow growth of the stable and cyclical fields predominated. More recently the doubling time has been much faster, and this suggests that at last the number of papers in dynamic fields is predominating just as it began to do in physics about the same

---

23. Sedimentary structures (0.0%): sedimentary structures
24. Tectonics (0.7%): tectonics or orogeny

Several categories were identified in the analysis of 1,423 papers but not pursued further as subfields, namely:

Biographies and history (4.1%)
Bibliographies (1.3%)
Annual reports, proceedings (3.2%)
Education (1.4%)
Engineering — ground water (3.6%)
Mineralogy — crystallography (6.1%)

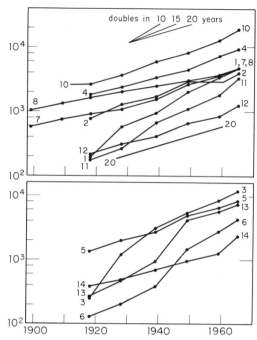

Fig. 3.7 *Above,* slow growth of stable subfields in the earth sciences. *Below,* variable growth in cyclical subfields. For key to numbers, see note 12 above.

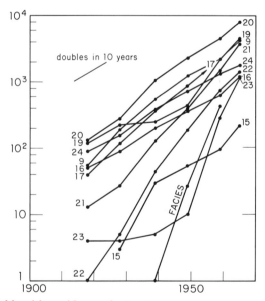

Fig. 3.8 Subfields with rapid growth. For key to numbers, see note 12 above.

time. Meanwhile, the careers of the average specialists in various subfields probably have been very different, even though all are scientists of one sort or another.

We now know that sciences grow at variable rates, but we have not considered why. We shall be in a better position to do so after evaluating some of the factors involved.

4 Population and Other Factors Affecting Growth

Science and technology include people employed as administrators, applied researchers, basic researchers, design engineers, development engineers, instrumentalists, laboratory assistants, specialists, students, teachers, technicians, and testers, to name but a few. Individuals are sometimes one and sometimes another, and, in addition, the classifications have indefinite boundaries. Naturally it is difficult to prepare a census of scientists or any other category, but we need it as a measure of the growth of science.

The United States Census provides comprehensive data of some categories as the census takers visualized them during the first half of the century.[1] In 1900, there were 38,000 professional engineers, 9,000 chemists, and 12,000 other scientists constituting 0.27 percent of the labor force (Fig. 4.1). After the National Science Foundation was established, it was charged with conducting a census of science, and it did so retroactively to 1940, thereby overlapping the census data.[2] The NSF identified the same number of engineers as the census in 1940 and 1950, but it found only 145,000 scientists compared with 210,000 identified by the census. In part this was because of a sharpening of the definition of "scientist" and the identification of 300,000 technicians, some of whom presumably regarded themselves as scientists when they filled out the census forms.

Even with the more restrictive definition, the number of scientists and engineers grew far more rapidly than the general labor force

[1] U.S. Bureau of the Census, 1958 and annually, *Historical statistics of the United States, colonial times to 1957, and continuations* (Washington, Government Printing Office).

[2] National Science Foundation, 1964, *13th Annual Report*, NSF 64-1 (Washington, Government Printing Office).

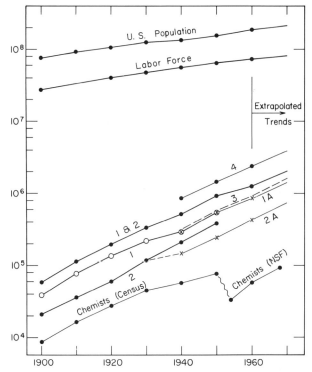

Fig. 4.1 Science and technology components of the general population and labor force. Professional (1) Engineers (U.S. Bureau of the Census, *Historical Statistics*); (2) Scientists (U.S. Bureau of the Census, *Historical Statistics*); (1A) Engineers (NSF 64-1); (2A) Scientists (NSF 64-1); (3) Technicians (NSF 64-1); (4) Manpower in science and technology (NSF 64-1).

to an estimated population of slightly over 2,000,000 in 1970. This was about 2.5 percent of the labor force, a tenfold increase in 70 years. Including technicians, the science and technology component of the labor force at present is 3,900,000 or pushing 5 percent of the whole.

Inasmuch as chemists were numerous enough to be a separate category in the U.S. Census, their number can be followed from 1900, when it was 9,000, to 1950, when it was 77,000, which suggests a slightly slower growth rate than other sciences. As with the whole population of scientists, it is difficult to compare the NSF census of chemists with the U.S. Census data, because in 1954 the NSF found 44,000 fewer chemists than the census did in 1950. Apparently most

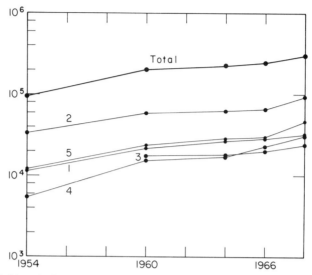

Fig. 4.2 A historical census of American scientists with some specialties differentiated. (1) Physics and astronomy; (2) Chemistry and chemical engineering when listed; (3) Earth sciences; (4) Mathematics; (5) Biological sciences.

of them were technicians (Fig. 4.1). The NSF census of individual sciences is available from 1954 through 1968 (Fig. 4.2).[3] It appears that natural scientists in all fields increased twofold in the period from 1954 to 1960, and then in the next 6-year period, the increase was much smaller. It is not certain that the increase before 1960 was really so rapid. The NSF sensibly makes a census by polling the members of scientific societies and analyzing the data returned. In 1954, if memory serves, many scientists did not bother to return the census forms, seeing little point in them. By 1960, more people were used to the NSF as being something different from the run-of-the-mill bureaucracy, and they returned the forms. Thus, the apparent growth may merely reflect sampling procedures. This might account for some of the great drop in the apparent number of scientists in general and chemists in particular between the counts of the census and of the opening efforts of the NSF. It is uncertain, consequently, whether or not the population of scientists increased any faster in the period 1954–1966 than the average rate for other decades in the century. From 1966 to 1968, there is a definite upturn compared with the average, but whether it will long persist remains to be seen. Since

[3] National Science Foundation, 1957, 1962, 1970, *American science manpower*, NSF 57-23, 62-43, 70-5 (Washington, Government Printing Office).

Population and Other Factors Affecting Growth

1960 it is possible, however, to distinguish somewhat different and probably real growth trends among the sciences. Biology and mathematics have almost doubled; this is reasonable considering the massive influx of federal money into medically related research and the enormous expansion of the computer sciences. Chemists, geologists, and physicists, in contrast, have increased at a relatively normal rate during the period.

As might be expected, the subfields of these sciences have grown at very different rates and with little regard for the average rates for the whole sciences. Physics has grown at a normal rate, but the population engaged in acoustics has dropped, and in atomic and molecular physics it has increased rapidly. Organic and analytical chemists have increased at a normal rate, but inorganic chemists decreased from 1964 to 1966 according to the NSF census (Table 4.1). Biologi-

TABLE 4.1. Census of Some Subfields in Science

Subfield	1960		1964	1966
Anatomy	1,130		1,010	1,030
Microbiology	4,530		2,790	2,950
Zoology	2,374		2,790	2,395
Ecology	900		1,150	1,350
Botany	1,710		2,240	2,130
Analytical chemistry	8,270		8,880	9,970
Inorganic chemistry	3,850		5,190	3,990
Organic chemistry	22,660		24,160	24,230
Chemistry (all others)	2,680	Biochemistry	6,340	7,500
		Physical chemistry	7,940	8,360
Acoustics	1,260		1,380	1,260
Atomic and molecular physics	840		1,770	2,060
Nuclear physics	3,790		3,170	3,560
		Particle	1,350	1,830
Optics	1,690		2,370	2,620
Solid state	3,140		4,150	4,590
Mathematics (other)	730	Other	450	810
		Numerical methods	5,730	8,470
Topology	420		640	810
Geometry	750		800	880

Source: National Science Foundation, 1957, 1962, 1970, *American Science manpower*, NSF 57-23, 62-43, 70-5 (Washington, Government Printing Office).

cal sciences have increased relatively rapidly, but somehow the numbers of anatomists, microbiologists, and zoologists have remained constant or even decreased. Mathematics has also increased rapidly, but specialists in such established subfields as geometry have not. The increase in mathematics occurred almost wholly in numerical methods as a consequence of the explosion in computer programming. In 1960, the census identified 730 mathematicians as working in the minor subfields grouped in the category "other." From this emerged 5,730 specialists in numerical methods in 1964 and 8,470 in 1966.

Once again it is evident that the average growth of science is the sum of some highly variable trends. Feast and famine may exist side by side. In the sense that scientists and students are not generally hungry and may be unaware of just what is happening, it may be more accurate to visualize some of the sciences and subfields as suffering from vitamin deficiencies. The effects can continue for a long time without the general weakening becoming apparent. The NSF census is some help in this regard but, perhaps inevitably, it suffers from the inherent difficulties of categorization. Consequently, we shall attempt a special one in geology, the science with which I am most familiar, but it will not be easy or necessarily definitive.

A historical census of American geologists can be based on the membership of geological societies, bibliographies of papers in geology, or, with less confidence, on the list of professional employees of government surveys or the number of degrees granted in geology. Of these measures the one with the most comparable results and the one with the longest continuous series of observations is the census of authors of papers. For the present purpose it is merely necessary to tabulate the year in which each author appears for the first time in the many volumes of the *Bibliography of North American Geology* of the Geological Survey.[4]

---

[4] The first cumulative bibliography and index, Bulletins 746 and 747, cover the period from 1785 to 1918 and are particularly valuable. The bibliography is not ideal for this census, because it includes geologists of the world who write about North American geology; but this does not seem to introduce much error. Compared to bibliographies in more basic sciences, it is relatively pure in that it includes few papers that would also appear in a bibliography in another science. There is the additional consideration that geologists have always taken "America" literally, and both Canadian and Mexican geologists and geology are indexed with those of the United States. Once again this does not seem to introduce much error in the census of the far more numerous geologists of the United States.

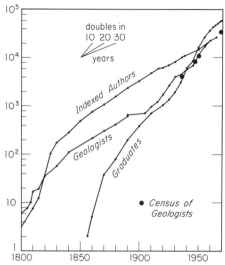

Fig. 4.3 Growth of the population of geologists and graduates with geology majors. "Geology" includes all types of earth science.

The first author listed in the bibliography published in 1785, the second in 1793, and the number rose rapidly to about 100 in 1825. The growth rate then slowed, but there were 500 authors by 1850. At that time, the growth stabilized into a simple exponential expansion with a doubling time of 18 years that continued until 1915, when the cumulative total reached almost 6,000 (Fig. 4.3).[5]

In 1915, this persistent growth trend slowed to an average doubling time of about 26 years until roughly 1950. The rate for the first third of this period was particularly slow. During the decade of the 1950's the doubling time returned to 18 years, but the sampling method is incapable of showing whether it remains the same now.

In sum, the historical census of new authors shows about the same changes in growth rate as the tabulation of pages of earth science publications. To a certain extent this would be expected, because the pages are published by the people, and more of one implies more of the other. The census is not of the number of geologists publishing at a given time, however, but of the number who began publishing up to that time. Most of the new authors, as will be shown later,

[5] Miss Constance Buffington had the stamina to tabulate 11,400 authors by year of first appearance for the period from 1785 to 1939. For the next two decades the number of new authors was estimated by decades from a sample of about 10 percent of the whole bibliography.

published one paper and vanished from the bibliographies. The major contribution has always been from a relatively small proportion of highly prolific authors. Thus, the counts of pages and new authors are related only to the extent that a constant proportion of new authors become prolific authors. This is more difficult to determine, but it appears that the proportion of prolific authors varies with circumstances, and the proportion was relatively low during the period of slow growth.

Nevertheless, the new-author census can hardly be taken as an independent confirmation of the variable growth rates indicated by page counts. Thus, we are driven to far more complex types of census in which none of the measures extends through the whole period of interest, and several must be spliced together. One method of making a historical census is to identify everyone working, writing, or teaching as a geologist in America during the time before regular geological societies came into existence. Although painful, this can be done by indexing every individual and the date on which his presence is first noted in Merrill's great volumes on the history of American geology. This procedure requires the usual sorts of arbitrary assumptions. I include people identified as topographic surveyors but not as ornithologists, unless I recognize the name as that of a paleontologist. I include a hodgepodge of summer staff, although rife with nepotism, because they were certainly working as field assistants when listed, and many became professionals. I exclude a small number of state survey directors and the like who were obvious political appointees and, moreover, incompetent. There were also some problems with distinguished foreign geologists who appeared briefly on the scene when already established. This small number was counted if they were active and not just tourists or lecturers.

The first 30 names indexed in this manner contain many who achieved enduring professional reputations: Silliman, Maclure, two Danas, Edward Hitchcock, Amos Eaton. Most, however, have become obscure like their successors. Their number mounted rapidly and reached 100 in 1840 (Fig. 4.3). From then until 1895, the number grew exponentially, with a doubling time of 20 years. It may have continued longer, but Merrill's volumes are comprehensive only to the end of the nineteenth century and, anyway, it did not seem profitable to continue this approach beyond 670 names. A check on the comprehensiveness of the tabulation is given by cross-referencing it with the list of early members of the Geological Society of America. In 1890, the GSA membership was 300, and it included 36 people not listed among the 600 identified by that time in Merrill's books. This seems curious, but in any event, it suggests the historical census is not in error by more than a few percent.

We must now turn to another measure in order to continue our census — the membership of professional geological societies. This is an obvious approach, but one difficult to follow. The first enduring scientific society in the United States is the American Philosophical Society founded in Philadelphia in 1743. The society merged with another local group in 1769, and Benjamin Franklin, a guiding spirit, was president from that time to his death in 1790. All aspects of science were of interest to the members, but geology was prominent. Thomas Jefferson was president of the society from 1797 to 1814, and during part of that time he was also President of the United States. He was an active vertebrate paleontologist both in Philadelphia and later in the White House.[6] The founders and early leaders of the republic had a much higher personal interest in and knowledge of science than our present leaders. Thus, it is hardly surprising that John Adams encouraged the founding of the second, still-existing, scientific society, the American Academy of Arts and Sciences, in Boston in 1780. John Quincy Adams, following Jefferson in more than one way, was simultaneously President of the United States and of this society in the 1820's. These were not mere local societies; Jefferson and Franklin, among others, were members of both. We could continue with this sketch of the history of learned societies with geological leanings, but with regard to a census the principal problem is already evident. Geologists, like all other men, tend to join more than one society. It is easy to obtain the number of members at any time in a professional society from its proceedings. Were there no overlap, the memberships could be summed to give a census. With overlap the individual members must be identified, a far more complex chore, but one that is done from time to time. In the early nineteenth century, moreover, it is always difficult to say just who should be counted as a geologist in a time when little specialization existed.

[6] G. P. Merrill, 1924, *The first one hundred years of American geology* (New Haven, Yale University Press; facsimile reprint, 1964, New York, Hafner Publishing Co.), p. 16.

The first society in which all the members were at least interested in geology was the Association of American Geologists, organized in 1840 in Philadelphia.[7] Unfortunately it was broadened to the Association of American Geologists and Naturalists in 1842, and by 1848 it had further expanded to become the American Association for the Advancement of Science. Consequently, the first national geological society that still persists is the Geological Society of America, founded in 1888. Its membership was 300 in 1890, and it grew only very slowly until 1945, when it was 950. Many other national geological societies were founded during that time. Fairbanks went to the trouble of sorting out all the individuals in 1936 and came to a total of 4,087 different members of geological societies in the United States.[8] Various surveys and estimates by the American Geological Institute Committee on Geological Personnel give a population of 8,500 in 1947 and of 11,000 in 1950.[9] For the last twenty years estimates of a geological census have been prepared by the American Geological Institute using the *National Register of Scientific and Technical Personnel* of the National Science Foundation, a source which we already know may be without peer but is not above reproach. The latest study gives 23,500 earth scientists in 1968 who respond to questionnaires about their occupations. Many people, of course, do not respond to questionnaires, and others give unreliable information. In 1969, the American Geological Institute identified 34,250 geologists as members of professional societies. This is the individual subscriber list of *Geotimes* and is far less than the sum of overlapping memberships. Presumably it includes few duplicate memberships and approximates the population of geologists. If so, about a third of them are not considered in the National Register.

The widely separated censuses can be used to calibrate the overlap of memberships in geological societies. These indicate that in 1935 the average society member belonged to 1.3 societies and in 1950 and 1965 he belonged to 1.7. The total membership of the major national geological societies has been accordingly adjusted to determine the increment of new society members at 5-year intervals. This increment can then be added to the census derived from Merrill's

[7] K. F. Mather, 1959, "Geology, geologists, and the AAAS," *Science 129*, 1106–1111.
[8] H. R. Fairbanks, 1936, "Geologists: their distribution and background," *Proceedings, Geological Society of America for 1935*, pp. 443–468.
[9] "Supply and demand for geologists, 1949–1950," 1950, *Bulletin of the American Association of Petroleum Geologists 34*, 1934–1942.

Population and Other Factors Affecting Growth

books up to 1890. This procedure indicates a lull in the growth of the geological population from 1890 to 1905, but this may be no more than the effect of joining the two censuses together. By 1910, the population began to expand rapidly with a doubling time of about 10 years until 1930. Then there was another lull for about 15 years and another period of rapid expansion until 1960. At that time, a third lull began and apparently still continues.

LOCUS AND TYPE OF EMPLOYMENT

Are all the identified scientists busy doing basic research and thereby writing all the identified scientific papers? Far from it (Table 4.2).[10] Scientists per se, excluding technicians and engineers, work at research and development including basic research, management and administration, teaching, production, and other minor occupations, and some are unemployed. The various NSF censuses are not entirely consistent, but most of the categories are identified from 1954 on. Research and development scientists almost tripled during the period from 36,700 to 96,000, but the fraction of scientists so engaged declined from 49 percent to 32 percent. Basic researchers were identified as such in 1960 when there were 31,100, and they have increased to 46,200 while remaining a constant 15–16 percent of the whole. Thus, it appears that basic research is becoming an ever-larger component of research and development. Scientists engaged in management and administration have multiplied almost fivefold since 1954, but their proportion has remained relatively constant at about 20 percent. Numbers of teachers have also grown about fivefold, and they have increased from 15–16 percent to more than 20 percent. Production scientists have likewise vastly increased, but at the same rate as the population of scientists. In sum, scientists are relatively evenly divided among those doing (1) basic research, (2) applied research and development, (3) administration, (4) teaching, and (5) other things including idleness; and, except possibly for teaching, the proportions have been relatively constant despite a great increase in population.

Scientists engaged in these various activities are employed in equally varied organizations (Table 4.3). The fraction of scientists

[10] National Science Foundation, 1957, 1962, 1970, *American science manpower*, NSF 57-23, 62-43, 70-5.

TABLE 4.2. Activities of Scientists

Activity	1954 Scientists (1000's)	1954 Per-cent	1960 Scientists (1000's)	1960 Per-cent	1964 Scientists (1000's)	1964 Per-cent	1966 Scientists (1000's)	1966 Per-cent	1968 Scientists (1000's)	1968 Per-cent
Total	74.2	100	201.3	100	223.9	100	242.8	100	300.0	100
Research and development total	36.7	49	74.9	37	77.7	35	80.0	33	96.0	32
Basic research	—	—	31.1	15	35.8	16	38.8	16	46.2	15
Management and administration	13.2	18	48.9	24	46.3	21	48.5	20	62.9	21
Teaching	11.7	16	29.5	15	41.2	18	43.5	18	62.1	21
Production	2.6	4	13.9	7	16.6	7	17.0	7	16.8	6
Other (including no report)	10.0	13	34.0	17	26.3	12	50.8	21	?	?
Unemployed	—	—	—	—	9.6	4	—	—	12.7	4

Source: National Science Foundation, 1957, 1962, 1970, *American science manpower*, NSF 57-23, 62-43, 70-5 (Washington, Government Printing Office).

TABLE 4.3. Employment of Scientists

Field	1954 Scientists (1000's)	1954 Per-cent	1960 Scientists (1000's)	1960 Per-cent	1964 Scientists (1000's)	1964 Per-cent	1966 Scientists (1000's)	1966 Per-cent	1968 Scientists (1000's)	1968 Per-cent
Total	86.5	100	201.3	100	223.9	100	242.8	100	300	100
Educational institutions	27.3	31.6	55.7	27.6	77.7	34.7	87.3	36.0	117.8	39.4
Federal government (total)	15.2	17.6	26.4	13.2	28.9	12.9	30.6	12.7	37.2	12.4
Military and Public Health Service	—	—	4.8		5.5		5.9		7.2	
Other federal	—	—	21.6		23.4		24.7		30.0	
Other govt.	—	—	10.7	5.3	7.5	3.3	8.3	3.4	10.0	3.3
Nonprofit organization	—	—	8.9	4.4	8.7	3.9	9.8	4.0	11.2	3.7
Industry, business, and self-employed	43.9	50.8	91.0	45.2	88.7	39.6	88.9	35.7	102.2	34.2
Other (including no report)	—	—	8.7	4.3	12.3	5.5	3.0	1.2	19.8	6.6

Source: National Science Foundation, 1957, 1962, 1970, *American science manpower,* NSF 57-23, 62-43, 70-5 (Washington, Government Printing Office).

in types of organizations has changed much more than their activities. Thus, in 1954, about 51 percent worked for industry, 32 percent for universities, and 18 percent for the federal government. Now the proportions stand as follows: industry, 34 percent; universities, 39 percent; federal, 12 percent; and the remainder are widely dispersed. The numbers of people in each of the major categories have expanded greatly, but proportionately the shift has consistently been away from industry and toward universities.

A comparison of activities and locus of employment is puzzling. The census consistently indicates that educational institutions include about twice as many scientists as teachers of science. The universities include no one in production and, simply because of the immediacy effect, the fraction of scientists old enough to be deans or even department chairmen cannot be very large. Thus, the nonteachers are largely occupied in research, and, at least in proper universities, it is basic research. However, if we subtract all teachers from the total number of scientists in universities, we approximate all basic researchers at any given time. Is there no basic research in industry? Are the Nobel Prizes awarded to scientists at Bell Laboratories a mistake? Or is it the census? Apparently only about one-third to one-half of the scientists at any given time participated in the census (Fig. 4.1, Tables 4.2, 4.3), and thus only generalities may be warranted on some subjects. The sample was very large, but why did some people participate and others not?

SCIENTISTS AS PAPER PRODUCERS

It is useful to have some measure of the average annual output of scientific papers per scientist. A casual inspection of the indexes of the various abstracting journals indicates wide variations in productivity, and this is confirmed by every detailed study. It is customary on academic appointment and promotion committees to judge three papers per year as an average output expectable in the University of California and presumably elsewhere. However, this is a highly select population with regard to writing papers, and the average for all scientists appears to be far less. This can be established by comparing our historical census of scientists with the output of scientific papers.

The most complete and consistent data at our disposal are those

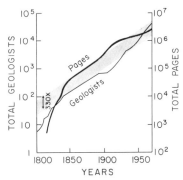

Fig. 4.4 Comparison of number of geologists and number of pages of geological reports.

comparing the population of geologists for 150 years with the number of pages of geological reports they produced. It is reasonable to assume 1 paper for each 10 pages. The years 1800 to 1840 were a period of rapid change, but with a gradual increase in the total number of pages compared to the cumulative total of geologists (Fig. 4.4). About 1840, the pages and geologists entered into parallel growth that persisted without significant change for 60 years and did not vary as much as twofold for almost a century. During that very long period, the ratio of cumulative pages to cumulative geologists was about 300–400. This means that on average each geologist produced that number of pages during his professional career, or 8–10 pages per year, or 1 paper per year, which seems reasonable. About 1920, when output began to approach a steady state, a remarkable change began. The number of geologists grew rapidly, but their output of pages did not; by 1955, the ratio had dropped to 100, or only one paper each 3–4 years. The average output for a geological career during the dormant period had dropped to only one-third or one-quarter that of the previous century. The rate of change is decreasing since the rejuvenation of geology, but the average output is now down to 60 pages, which is very few equivalent papers in a whole career. We may think that the many new earth scientists have not yet had a chance to write their 300 pages, but that was true of the previous century as well. The cause of the collapse of productivity is elsewhere.

We can look to other sciences for other indications of the productivity of scientists as paper writers. A long series is available in chemistry, because chemists are identified in the U.S. Census from 1900, and the annual output of papers by *American* chemists has been abstracted from *Chemical Abstracts* for selected years.[11] In 1900, there were 9,000 chemists who wrote 3,000 scientific papers, an average rate of 1 in 3 years instead of 3 in 1 year. By 1930, 45,000 chemists wrote about 11,300 papers, and in 1950, the numbers were 77,000 chemists and 19,400 papers. We know that this census may include technicians, but it appears that the trend in geology is confirmed. Not only do chemists identified in this way write only one paper in 3–4 years, but also the population is increasing faster than the output. The data on American papers in chemistry for 1951 and 1960 overlap the more selective census of chemists by the National Science Foundation. By this definition, 28,000 chemists in 1953 wrote 22,000 papers or approaching 0.8 papers each. The same data give 52,000 chemists writing 31,000 papers in 1959, or an average rate of 0.6 papers per year. We know the early NSF census may be less complete than the later ones, which could account for some of this change. Nevertheless, it appears that chemists write few papers on average, and the output is decreasing with time as it is in geology.

However, during the last few years the annual output of geology has at last begun to increase for the first time since 1920. Is that also occurring in chemistry? It appears so. If the American component of world chemistry remained constant at the 25 percent found in 1960, and the extrapolated output was 45,000 papers in 1965, the NSF census gives 66,000 chemists who thus wrote 0.7 papers each.

The NSF data on physics and its subfields also confirm an increase in average output and at least imply that this is only a recent phenomenon. The central problem confronting the analysis is that the population is for American physicists, and the output of papers in *Physics Abstracts* is for the whole world. For physics as a whole the world output of papers in 1954 and in 1960 was just about equal to the population of American physicists (Table 4.4). Thus, all physicists in the world averaged less than 1 paper per year. The average output compared to American physicists then began a steady rise that still continues. An increase in the fraction of the total output attributable to foreign physicists might account for part of this, but a consideration of subfields suggests that most of it is a real increase in American output.

---

[11] National Academy of Sciences, 1965, *Chemistry: opportunities and needs*, Publication 1292 (Washington, National Academy of Sciences). Commonly called the Westheimer Report. Papers and patents are both discussed here as "papers."

TABLE 4.4. Population of American Physicists Compared with World Output of Papers in Physics and Its Subfields

Year	Physics		Acoustics		Optics		Atomic-Nuclear-Particle		Solid State	
	People	Papers	People	Papers	People	Papers	People	Papers	People	Papers
1954	11,200	11,500								
1960	20,900	21,500	1,260	590	1,690	700	4,630	6,300	3,140	5,300
1964	26,700	32,200	1,380	720	2,370	960	5,290	9,300	4,150	9,500
1966	29,100	36,000	1,260	800	2,620	1,150	7,450	11,400	4,590	12,500
1968	32,500	51,000								

Source: National Science Foundation, 1957, 1962, 1970, *American science manpower*, NSF 57-23, 62-43, 70-5 (Washington, Government Printing Office); National Academy of Sciences, 1966, *Physics: survey and outlook* (Washington, National Academy of Sciences).

In acoustics and optics the total world output is about 1 paper for each 2 American specialists in the subfields, and the average productivity is about the same as in American geology. These ancient and slowly growing subfields dominated physics output until fairly recently, and presumably they indicate the general level of output earlier in the century. In the newer giants of nuclear and solid state physics, the output ratios and apparent growth are very different from the older ones. World output in nuclear physics papers is increasing rapidly but maintains a relatively constant ratio of 1½ per American specialist. Perhaps in the whole world the average output is 1 paper per year or much faster than in the old subfields. In solid state physics not only is the output ratio large, but it increased from 1.7 in 1960 to 2.3 in 1964, and to 2.7 in 1966. This is a subfield dominated by the United States which "has unique strength." [12] If most of the research is therefore done by American specialists, they have a remarkable record of average output quite unlike any other identified for a large group of scientists. Indeed, with the increased output of solid state physics removed, the output in all other subfields of physics has hardly changed at all. It was one world paper for each American physicist in 1954 and remained so as late as 1966. Considering an undoubted increase in the population of physicists in the world, this means that except for one subfield, the output of physicists probably has declined.

Viewing the data dispassionately it appears that, for most of this century, output of geology has been relatively independent of the number of geologists. The population increases come what may. Fragmentary information suggests that this may also be true of physics and chemistry. Thus, the exponential growth of the number of scientists appears to be merely an example of Parkinson's Law on the growth of bureaucracies. We can only conclude that the number of scientists will continue to increase whether or not any more papers are published at all. Perhaps that is too strong. Some papers will always appear as long as publication is a criterion for promotion. Price has expressed concern that the distribution of intelligence in the general population will limit the number of scientists. The relationship we have been examining suggests that "scientists" will be found to fill the appropriate slots in tables of organization. At

---

[12] National Academy of Sciences, 1966, *Physics: survey and outlook*, Publication 1225 (Washington, National Academy of Sciences), p. 21. Commonly called the Pake Report.

Population and Other Factors Affecting Growth

present, less than one-fifth of scientists are engaged in basic research, and they surely publish more papers than average. If even a small number of them shift to other activities, the research output will plummet. The limit is not on scientists taken as input to a social system, but rather on science taken as output.

CAUSES OF VARIATIONS IN GROWTH OF SCIENCE

Federal bureaucrats and those who advise them tend to voice a simplistic cry that if federal funds for science are not constantly increased, the output of science will suffer. Occasionally this is put in the more positive form, that if there are more funds, science will prosper. It is often difficult to be heard without shouting in Washington, and it is hard to shout anything but a simple message. Nevertheless, this attitude fosters much of the concern about the future of American science, and it seems worthwhile to examine and develop it.

In order to produce research, a scientist needs freedom to do it, a salary so he can eat, support for his work, new equipment or a new paradigm, and new ideas. Exploration of new regions and lengthening time series provide yet other sources of stimulation in environmental sciences such as astronomy, geology, ecology, and oceanography, because the probability of discovering new types of features increases. All of these variables are affected by money in the sense that a scientist cannot do research if he is not eating. Beyond that, however, the influence of funding or at least federal funding is less obvious.

Let us examine federal funding for research and development (Fig. 4.5).[13] The government has supported science since the dawn of the republic with resulting contributions from the Coast and Geodetic Survey, Geological Survey, Naval Observatory, Department of Agriculture, and so on.[14] We have no comparable data on the population of scientists or the output of papers except for the gen-

[13] National Science Foundation, 1965, *Federal funds for research, development, and other scientific activities*, NSF 65-19 (Washington, Government Printing Office). This is a recent example of a continuing series. V. Bush, 1945, *Science the endless frontier*, National Science Foundation, NSF 60-40 (Washington, Government Printing Office), reprinted 1960; P. M. Boffey, L. J. Carter, and A. Hamilton, 1970, "Nixon budget: Science funding remains tight," *Science 167*, 845-848.
[14] Treasury Department, 1874, et seq., *Digest of appropriations for the support of the government of the United States* (Washington, Government Printing Office).

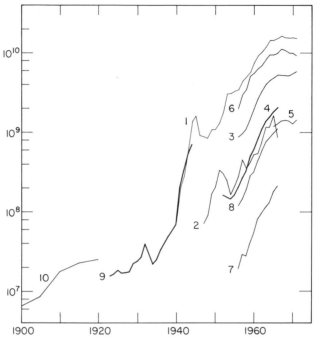

Fig. 4.5 Federal funding for research and development by fiscal years (NSF 65-19; *Science 167*, 845–848). Federal expenditure for: (1) Total research and development; (2) Research and development plant; (3) Research; (4) Basic research; (5) Research in colleges and universities; (6) Development; (7) Basic research in biology; (8) Basic research sciences; (9) Government research (NSF 60-40); (10) Environmental agencies plus other research (*Digest of Appropriations*).

eralization that science on average doubles every 15 years. Nevertheless, the early history of federal science and science support deserves our concern as a guide to the origin and development of all that followed.

Support for scientific research was rarely identified as such in the Treasury Department *Digest of Appropriations* in the nineteenth century. I have dredged what I can from the *Digests* at 5-year intervals, starting in fiscal year 1875 when they began. The Bureau of the Budget came half a century later, and the original records have a quaint charm. Much of the *Digest* in 1875 was concerned with "Indian Affairs" within the United States and "Foreign Intercourse" elsewhere. Meanwhile, or already, the Corps of Engineers was busy "Improving X River." I do not include this as research. However, the War Department was engaged in scientific research in 1875 in "experiments with breech loading cannon" ($20,000), and in 1890 with weapons experiments ($15,000) which included work on "torpedo howitzers." By 1905, it had established a proving ground for experiments and development. The record of the Navy, then as now, seems both broader and more distinguished. In 1875, it spent $30,000 for construction and operation of facilities at the Naval Observatory and another $240,000 for the same purpose only 15 years later. These and other facilities were used for "discovery of new planets" and to observe the transit of Venus, thereby following the precedent of Captain Cook a century earlier. The new observations required allotments "for pay of computers," and they may also have been of use in the famous "experiments on the velocity of light" that appear in the budget in 1880. All these expenses were dwarfed by the costs of polar exploration which, as a latter-day explorer, I count as research. I do not know whether the nation contributed to the tragic explorations of Sir John Franklin or to the voyage of the *Fox* searching for him. It was involved in an expedition to Lady Franklin Bay, named after Sir John's wife, and in 1885 it spent $774,000 for relief of that expedition and paid an additional reward of $25,000 for the rescue. Oceanography has always come dear.[15] By 1910, the Navy was spending $100,000 for weapons experiments and had an experimental station at Annapolis which cost $25,000 to operate. A decade later Navy costs for experiments and development, identified as such in the budget, were $400,000, just the same as the cost of the Army proving ground.

Meanwhile, various Federal agencies were established to deal with environmental problems, and they too were doing some research as well as surveying and monitoring.[16] Just how much is "research" depends on the definition, but in 1875 it probably included "inquiry respecting food fishes" in the Treasury Department, the operation of the Smithsonian Museum, and the Geological Survey of the Territories in the Interior Department. Most of the civil research budget, however, was in the Department of Agriculture, which was engaged only in research, operation of laboratories and experimental gardens, and supporting operations. It alone spent $308,000, which supported nine scientists in their work.

In 1875 the total military budget, for what is here identified as

[15] H. W. Menard, 1969, *Anatomy of an expedition* (New York, McGraw-Hill), Chap. 1.
[16] See Chapter 8.

research, was $112,000, and the civil budget was $494,000, of which most was agriculture. These general relationships hardly changed during the next 45 years. The pattern that emerges is that after each environmental agency was founded it grew rapidly to a modest size, and after that most of its work hardly qualified as research. Thus it seems likely that research was important when the Office of Standard Weights and Measures spent $10,900 in 1900, but not so much so when the newly created Bureau of Standards spent $174,000 5 years later. Likewise, the fish inquiries continued and smell like research until they were consolidated into a separate bureau. It seems fair to include the costs of building Woods Hole and the operations of the research ship *Albatross,* and I have included a "portrait of Joseph Henry" ($500) in 1880 and a "statue of Joseph Henry" ($900) in 1885 as items that would now be concealed under the research budget. At this point it will be clear that the process of identifying research in ancient records is subjective, but so is it now, and I shall make no further apologies.

Although the military were always engaged in research and environmental observations, the efforts were small compared to those of civilian agencies, and the proportion steadily decreased. This occurred in two ways. Military research generally decreased from 1875 to about 1900, when it was institutionalized in experimental stations and proving grounds. Then, of course, it went through a bureaucratic expansion. Meanwhile, civil agencies involved in research proliferated and grew, and the Department of Agriculture led all the rest. As they became bigger they also became bureaucratic, but they easily dominated federal research.

The other effect that prevented the early occurrence of the dominance of military science that now exists was perhaps more important. The geological surveys and the lake surveys and the observation and prediction of weather were originally responsibilities of the Army, presumably because it had logistics support and a telegraph service. After 20 or 30 years, most of these services were widely used and appreciated, and they were transferred to special civilian agencies. That is about the interval since the military assumed a dominant role in federal research during World War II. It should be the example for our times. Research support flourished under the Office of Naval Research, and less enlightened agencies and the administrators and civil servants involved deserve every praise, but now is the

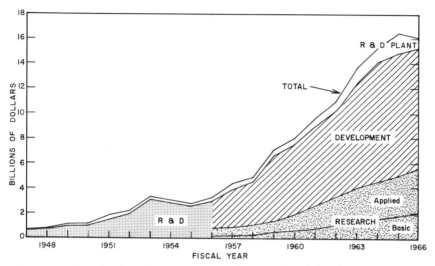

Fig. 4.6 Federal funding for components of research and development.

time to return government research and science to civilian control where it resided from 1776 to 1940. During that period science apparently expanded at an average rate not very different from now, and the information base for the present technology accumulated; a very credible record. Unfortunately, we cannot analyze it in detail for lack of data. This is not true of later years, when the record becomes increasingly clear.

Federal support for research and development in the twentieth century can be approximated from the reports of the National Science Foundation and before that from the *Digest of Appropriations for the Support of the Government of the United States* and various data from the Bureau of the Budget. The components, such as basic research, were identified at different times as they grew large enough to be of interest, and thus they constitute a haphazard record, but one that is nevertheless revealing (Figs. 4.5, 4.6). Development has been the overwhelming sink for federal money for science ever since it was first identified. Prior to that time, routine measurements such as weather observations and land surveying were dominant. Applied research has been next in importance when identified, and basic research and R and D plant have been tied for last.

If we compare federal funding for research and development since 1900 with the output of scientific papers (Fig. 4.7), it is evident

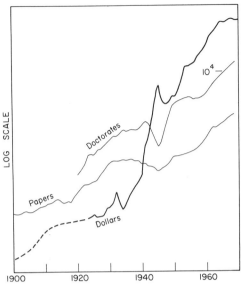

Fig. 4.7 Federal expenditures for research and development compared with output of scientific papers and doctorates in science and engineering.

that during the seventy years both increased. Otherwise the correlation is minimal. In the first two decades, expenditures increased and output was roughly constant. From 1920 to 1935, papers increased rapidly and dollars during the depression were roughly constant. In 1935, an enormous expansion of federal funds for research began and continued for a decade. During the same period the output of papers gradually declined. From 1950 to 1965, for the first time, both papers and budget increased rapidly. Has the golden age arrived? Hardly, because during the last 5 years expenditures have remained constant and the output of papers has steadily increased. This is a shaky attempt to correlate world output of scientific papers with U.S. federal budgets for science-related activities. The results are not at all reassuring. It seems that during much of the century indifferent government support has not prevented a rapid expansion of scientific papers. Likewise, an explosive increase in support has been correlated with a decline in output as often as with an expansion.

The correlation between research paper output and expenditures for basic research is better but it still leaves something to be desired (Fig. 4.8). Papers increased only slowly in the mid-fifties, when basic research was funded at a constant level. They increased rapidly in the

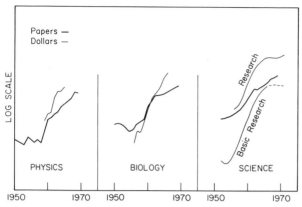

Fig. 4.8 Federal funding for basic research compared with output of science.

late fifties and early sixties at the same time as funds for basic research; and both began to slow about the same time. However, federal funds for all research and for university research varied little in the late sixties, and presumably this was true for basic research as well. Meanwhile, the exponential growth of scientific paper output did not even pause.

Individual sciences can be studied in the same way but for only brief periods. The output of papers in biology and physics shows a close correspondence with the rapid expansion of funds for basic research in the same sciences.[17] However, the papers seem to be what is called a "lead indicator" in economics. The slopes of the output curves decrease long before the supporting funds show any change in exponential growth.

In sum, it appears that money is helpful, but an increase does not long sustain a proportional expansion of science, and level funding does not quickly suppress growth. How long exponential growth will continue if basic research is level-funded is another matter. American scientists tend to have enormous accumulations of undigested data, rooms full of virtually idle analytical equipment, and other winter stores that will sustain research output at a much lower level of funding for a while. A sort of cottage industry of widespread small science could turn out a lot of papers and doubtless some of them would be very useful, but gradually science would change. The reason undigested data are so abundant is that after they were col-

---

[17] Funding for physics is taken from *Physics: survey and outlook,* the Pake Report.

lected they did not lead to anything that looked very important. Perhaps a new, hypersensitive, and extremely fast instrument made all existing measurements obsolete. Thus, the continuing mining of low-grade research ore may merely keep scientists employed and yield very little of value. The history of geology illustrates what can happen when information is abundant but the techniques available are inadequate to solve important problems.

Geology is the science of the history of the earth, and its development can be compared to that of the historical reconstruction of a long-dead civilization. The initial problem is to translate the language on the best of the ancient tablets. This was done in geology about 1800 with the discovery that layers of sedimentary rock contain fossils in a regular sequence and the realization that the younger layers are on top. Simultaneously, Hutton and Playfair were demonstrating that the physical and chemical processes now active were also active in the past. The next step in a historical reconstruction is to try to piece together and translate a vast pile of broken tablets and pottery. This occurred in geology in the period 1830–1860, when the existence and origin of tight folding and thrust faults were published along with fundamental concepts of mountain formation.[18] During this period the concepts of evolution and of a recent ice age also enormously improved the translation of the history of the earth.

With these concepts and techniques it might be possible to understand a simple civilization, but what if it was complex and had a long history? We now know that the geologists of the first half of the nineteenth century could not possibly reconstruct the origin and history of the earth, because they lacked any knowledge of most of geological time and of most of the surface of the earth and all of the interior. This realization must have gradually dawned on the more inquiring minds in geology. In our analogy, most of the history was in a second language that could not be translated with fossils, but required the discovery of isotopes and radioactive decay. Moreover, most of the tablets in both languages were in inaccessible regions, given the resources and technology of the times. Ocean-

[18] F. D. Adams, 1938, *The birth and development of the geological science* (reissued 1954, New York, Dover Press); A. Geikie, 1897, 2nd ed. 1905, *The founders of geology* (reissued 1962, New York, Dover Press); G. P. Merrill, 1924, *The first hundred years of American geology* (reprinted 1964, New York, Hafner Publishing Co.); H. B. Woodward, 1907, *The history of the Geological Society of London* (London, Geological Society of London).

Population and Other Factors Affecting Growth

ographic ships, geophysical instruments, and equipment for high-pressure laboratory simulation of the interior of the earth all remained to be developed.

Thus, in a most fundamental way, geology was barren of testable ideas. In 1915 Wegener proposed continental drift, and in 1923 Chamberlain proposed the planetesimal hypothesis of the origin of the earth, but the tools were lacking to test either, and they lay fallow. Otherwise, there was little to do except continue to use the old tools of translation to solve ever more trivial problems or else to speculate on what the untranslatable or unrecoverable records might someday reveal. This was the period when geology was relatively dormant. It has ended; new ideas and new tools abound in geology. But what happened to one science could happen to others, if the circumstances were the same.

5 Papers and Citations and Scientific Fame

A movie or television personality is in an occupation that may make him, or her, or part of her, famous in the sense that a sizable fraction of the population can recognize a picture or a name. Scientists, like other people, are not in such occupations, and consequently, as scientists, they do not become famous. Haberdashers and corporation lawyers are also in occupations that do not lead to fame in themselves. However, if an individual haberdasher becomes, for example, President of the United States, then he becomes famous. This can also happen to a scientist, but it has not. Having committed themselves by long and hard work to science, most scientists tend to stick with it. Consequently, they are little known to the general public.

Einstein is the only scientist I can think of who became really famous simply by doing science. Neils Bohr was famous as a scientist but more in Denmark than the world. Charles Darwin did hardly anything else but science, but he first became well known for his best-selling account of the voyage of the *Beagle*. Thomas Alva Edison certainly was a famous scientist, but the degree of his nonconformity to the normal practice of science is that he obtained 1,061 patents but wrote no scientific papers. I suspect that the most famous scientists are those professionally engaged in using mass media. J.-Y. Cousteau, whose achievements have brought him election to prestigious scientific societies, is one of the very few making movies and television serials. Thor Hyerdahl, likewise, is highly successful in writing about his scientific adventures, and his books are in every bookstore and every airport, although not in every living room like Cousteau's television show.

Papers and Citations and Scientific Fame

Far below such personalities in fame are the scientists who do something which makes them as well known as, let us say, the Secretary of State; a poll might show that 1–3 percent can be recognized by men in the street. These are scientists whose names and sometimes pictures appear frequently in print and for a prolonged period. Thus, they become as well known as the more exposed but more transient public officials. Scientists who found businesses and are listed in *Fortune* as having assets of $200–500 million may become well known if they do something with the money. Scientists who become president of Harvard, Ambassador to Germany, and then analyze public school education may reach such a measure of fame; likewise, all winners of two Nobel Prizes who take vigorous stands on political issues and march in parades. Former Special Assistants for Science and Technology to the President of the United States who take strong and open stands against the military-industrial complex may also become famous, although, if they all do so, the fame may go to the group rather than the individual.

Most of the best-known scientists are personalities only to the relatively limited public of fellow scientists. The Nobel Prizes in science get better publicity than the Man-of-the-Year-Award in Life Insurance, but six months later the recipients are identifiable mainly by scientists and insurance salesmen respectively. After all, the Miss America contest gets far more publicity than the Nobel award ceremony, and few can long remember the winner except people involved with advertising.

Thus, in assessing the fame of scientists and how to get it, we need merely consider their professional achievements in such things as the following: publishing papers, advancing the scientific research front, founding or leading a school of science, editing a technical journal, officiating in a professional society, or presiding over a science-oriented government agency. We can attempt to measure their success as scientists by whether they win Nobel or other major prizes or are elected to the more prestigious organizations such as the Royal Society, Akademiia Nauk, or the National Academy of Sciences. We can attempt to see how it was done by examining the papers they publish and the citations they receive — which is a measure of their contribution in advancing the research front.

Fig. 5.1 The Scientific Literature. Reprinted from C. E. Wegmann, 1939, "Zwei Bilder für das Arbeitszimmer eines Geologen," *Geologische Rundschau 30*, 390. See also F. Darwin, ed., *The life and letters of Charles Darwin* (New York, D. Appleton & Co., 1897), 303, "I have long discovered that geologists never read each other's works, and that the only object in writing a book is a proof of earnestness, and that you do not form your opinions without undergoing labor of some kind."

Papers and Citations and Scientific Fame

PAPER OUTPUT AND SCIENTIFIC FAME

One way to identify a scientist is to mark him as a man who writes a paper published in a scientific journal or indexed with or listed among the abstracts of such journals. Scientists are in the paper-publishing business (Fig. 5.1). We can now address two questions. Is a prolific publisher a successful scientist? Is it necessary to be a prolific publisher to be a successful scientist? For convenience we define "success" as being elected to the National Academy of Sciences. We can identify prolific publishers in geology during the last 200 years by merely summing the listings in the various bibliographies of North American geology. The fifty most prolific to date are listed in Table 5.1.

The nineteenth-century British mathematician, Cayley, published 995 papers, which was formerly taken as the record for productivity.[1] We might pause a moment and contemplate 1,000 papers, an average of 1 every two or three weeks without interruption for 50 years. Each scientific paper requires some research and then generally a rather elaborate procedure with many steps before publication. It is prepared in rough draft, then in final draft with illustrations, then submitted to an editor, then sent to reviewers, reviewed, and returned. Then, some weeks or months after it was first written, the manuscript is back in the hands of the author, who meanwhile has written 1 or 2 other papers. He then attempts to modify the half-forgotten manuscript according to the comments of the reviewers and editor, and he resubmits it. After two to six months (or 2 years during the dormant period in geology) the manuscript and galley proofs reach the hands of the author. Meanwhile, once again, he has written 3 to 12 more papers and also revised the same number but not entirely the same series of manuscripts. The proofs are then checked and returned to the editor along with a request for reprints for separate distribution. In due course, the paper is published in a journal, and the editor sends the reprints, which are then distributed. The whole process requires that the author read each manuscript four times after the first draft. He sends or receives the paper

---

[1] Derek J. de Solla Price, 1963, *Little science, big science* (New York, Columbia University Press), p. 49. Professor Price has called my attention to a reported output of 3,904 papers by T. D. A. Cockerell, but Table 5.1 indicates that only 203 papers are indexed in the *Bibliography of North American Geology*.

TABLE 5.1. Prolific Earth Scientists

Name	Subfield	Total	1785–1918	1919–1928	1929–1939	1940–1949	1950–1959	1960–1969	Papers Cited†	Citations	Maximum Citations/ Paper
E. D. Cope*	vertebrate paleontology	1,395	1,395	–	–	–	–	–	42	45	3
C. R. Keyes	geology	1,293	320	294	505	174 (in 1940–1942)	–	–	1	1	1
J. A. Cushman	micro- paleontology	427	20	113	251	40	3	–	26	27	2
W. M. Davis*	geomorphology	387	320	37	29	1	–	–	18	27	6
James Hall*	invertebrate paleontology	360	360	–	–	–	–	–	7	7	1
H. F. Osborn*	vertebrate paleontology	330	240	50	39	1	–	–	13	16	4
E. W. Berry*	paleobotany	329	139	105	64	14	7	–	13	13	1
J. W. Dawson**	invertebrate paleontology	320	320	–	–	–	–	–	16	17	2
W. R. Tillson	geology	284+	3	108	42	46	52	33+	0	0	0
T. S. Hunt*	geochemistry	240	240	–	–	–	–	–	1	1	1
A. C. Lane**	geology	236	150	30	43	13	–	–	6	6	1
N. H. Darton**	geology	224	180	25	15	4	–	–	8	9	1
J. D. Dana*	geology	220	220	–	–	–	–	–	14	15	2
O. C. Marsh*	vertebrate paleontology	220	220	–	–	–	–	–	3	3	1
J. M. Clarke*	invertebrate paleontology	211	180	31	–	–	–	–	4	4	1
J. F. Kemp*	economic geology	208	180	28	–	–	–	–	2	2	1

Papers and Citations and Scientific Fame

Name	Field										
C. Schuchert*	stratigraphy	208	95	53/2	57/1	3/–	–/–	–/–	13/0	16/0	3/0
N. H. Winchell**	geology	203	200	2	1	–	–	–	0	0	0
T. D. A. Cockerell	invertebrate paleontology	203	132	42	24	5	–	–	20	20	1
T. W. Vaughn*	invertebrate paleontology	202	100	51	38	13	–	–	1	1	1
J. Leidy*	vertebrate paleontology	200	200	–	–	–	–	–	12	14	2
W. P. Blake	economic geology	200	200	–	–	–	–	–	2	2	1
J. S. Newberry*	geology	200	200	–	–	–	–	–	3	3	1
W. D. Matthew*	vertebrate paleontology	198	140	39	18	1	–	–	6	7	2
G. P. Merrill*	astrogeology	183	120	58	5	–	–	–	6	7	2
F. P. Shepard	marine geology	181+	0	13/42	59/2	32/–	43/–	34+/–	60/10	107/11	10/2
T. C. Chamberlin*	geology	182	140	42	–	–	–	–	10	–	–
G. F. Matthew	invertebrate paleontology	180	180	–	–	–	–	–	2	2	1
R. C. Moore**	invertebrate paleontology	179	4	41/12	49/–	47/–	28/–	10+/–	50/1	50/1	1/1
E. O. Hovey	vertebrate paleontology	172	160	–	–	–	–	–	–	–	–
G. G. Simpson*	paleontology	167+	0	27/45	48/40	49/–	25/–	18+/–	95/9	186/10	27/2
E. M. Kindle	sedimentation	165	70	13	9	–	–	–	3	4	2
C. A. Hollick	geology	162	140	–	–	–	–	–	–	–	–
C. A. White*	invertebrate paleontology	160	160	–	–	–	–	–	0	0	0
E. H. Sellards	geology	157	70	28/41	41/29	–/18	–/2	–	1	1	1
C. D. White*	paleobotany	152	80	41	29	2	–	–	0	0	0
O. P. Hay	vertebrate paleontology	151	96	47	8	–	–	–	4	4	1

TABLE 5.1 — Continued

Name	Subfield	Total	1785–1918	1919–1928	1929–1939	1940–1949	1950–1959	1960–1969	Papers Cited†	Citations	Maximum Citations/Paper
C. Stock*	vertebrate paleontology	145	0	26	80	31	8	—	2	2	1
C. Palache*	mineralogy	142	42	34	42	20	4	—	50	61	6
B. F. Howell	invertebrate paleontology	141+	0	10	45	51	29	6+	6	7	2
G. K. Gilbert*	geology	141	140	1	—	—	—	—	12	18	5
R. P. Whitfield	invertebrate paleontology	140	140	—	—	—	—	—	1	1	1
R. Ruedeman*	invertebrate paleontology	136	45	27	52	12	—	—	7	7	1
C. K. Wentworth	geology	135+	1	41	48	16	7	2+	20	22	3
E. Blackwelder*	geomorphology	132	36	26	48	18	4	—	13	16	3
F. W. Sardeson	geology	126	40	36	49	1	—	—	0	0	0
A. K. Miller	micro-paleontology	126	0	1	46	50	29	—	0	0	0
W. H. Twenhofel	sedimentation	123	14	34	40	27	8	—	17	23	4
E. B. Branson	geology	123	25	21	39	33	5	—	1	1	1
D. W. Johnson*	geomorphology	121	50	18	44	9	—	—	15	21	5

Source: U.S. Geological Survey, *Bibliography of North American Geology* (Washington, Government Printing Office). Summary volumes are issued each decade.
* Member NAS.
** President or medalist GSA.
† Science Citation Index 1968.

in its various forms through the mails a total of seven times. Thus, in any year, the writer of 1,000 papers is required to write 20 times, read 80 times, and attend to the post 140 times. That takes care of his own papers, after which he can read and review those by his colleagues.

In this perspective, the achievement of C. R. Keyes in writing 1,293 papers can only be viewed as phenomenal, not only as a scientific output but as an example of directed human energy. He published for 54 years, from 1888 to 1942. For the whole period of the 1930's he averaged 45 papers per year, and in the first 3 years of the 1940's he published 174 papers. If ever a scientist should have achieved honors and fame, he would appear to be the man; neither came to him.

This is not true of most of the other prolific producers of papers. Twenty-three of them wrote 200 papers or more, and of these 14 were elected to the National Academy of Sciences and 4 others were either presidents or medalists of the Geological Society of America. Three-fifths of the whole 50 received one of these particular honors, and almost all have received some major honor from the profession. There should be no doubt that the most prolific writers are much more likely to receive recognition than those who are less so.

The folklore equating number of papers with scientific achievement and recognition is widely known. Consequently, it is not unheard-of for scientists and erstwhile scientists to turn into paper mills pouring out a flux of pages, sometimes without much regard for content. To such men the endless procedure of rewriting, revising, mailing, and reading may pose unendurable limitations on output. A staff of laboratory assistants, secretaries, and mail clerks is almost essential for a very high producer in the sciences and probably in any other field. However, even such support does not give the optimum solution to minimizing effort and speeding publication, which can only be achieved by gaining control of a journal. It can hardly be a coincidence that each of the three most prolific men did exactly that. Keyes controlled and published almost entirely in the *Pan American Geologist,* Cope in the *American Naturalist,* and Cushman in the *Journal of the Cushman Foundation.* It is striking that two of them have not received any of the honors that I have examined. This is true also of W. R. Tillson, who is the only other person among the most voluminous writers who is in the same position. He publishes his work by a commercial press in Louisville. Thus, we can venture a generalization. A scientist will be highly honored if he publishes 100 or more papers, but only if he can do it without capturing a printing press.

A few characteristics of the group of prolific earth scientists appear in Table 5.1. The period of maximum productivity for most was prior to 1919, when some of the big producers wrote 20 papers a year. Presumably they had control of a journal or lived next door to the editor or had lots of assistants. Among the younger ones, 11 had a period of maximum productivity in the 1930's. The maximum average in recent years is about 5 papers per year. In terms of our model of the growth of the earth sciences, most of the most prolific producers were active in the early period of normal science and before dormancy began in geology. This can hardly be attributed to a change in energy of scientists or in availability of research problems. More likely it is related to the volume of scientific literature. The potentially prolific man in the nineteenth century did not have to read as many background references before starting to write up his research. By the early twentieth century, a potentially prolific man had to digest the output of the prolific men who came before him, and the situation has steadily worsened in the older subfields. It appears unlikely that many people will publish more than 200 papers in geology in the future, at least without gaining control of a printing press.

Almost half of the prolific writers are paleontologists, which is grossly disproportionate to their representation among earth scientists. It is relatively easy to write one type of paper in paleontology, namely, the description of a new species, and this may account for part of this high productivity. But an additional factor should not be neglected. There are already so many papers in all fields of paleontology that a sizable effort is necessary for a new scientist to achieve recognition and gain the esteem of his colleagues and his place in the queue.

We have found that prolific publication does not guarantee "success" as we have defined it, but it certainly helps. We now turn to the related question of whether it is necessary for success. With this in mind, we should remember that the productivity records for geologists are for a whole lifetime and are not those at the time of election to the National Academy. Cope, as a notable example, published 1,395 papers, but only 71 before election at the age of 32. Indeed, in geology there are the two extremes of William Morris Davis, who published 271 papers before election, and Alexander Agassiz, who published none that are indexed as geology.

A measure of the necessity for prolific paper production, although cruder than the number at election, is the total career output for members of the National Academy; this information can be derived from the *Biographical Memoirs*.[2]

Tables 5.2 and 5.3 make it clear that regardless of field, members of the National Academy tend to be prolific paper producers. The median for 67 members is 130 papers. Once again, Edward Drinker Cope leads all the rest with 1,395, but Alpheus Packard published 579 in zoology and Robert Ridgway 540 in ornithology.

Perhaps this prolific production was customary, but it certainly was not necessary. Henry Bumstead published only 18 papers, and three other physicists, E. F. Nichols, Leo Szilard (Table 5.3), and A. W. Wright wrote only 26, 30, and 32 papers respectively. Wright published only 20 listed by his biographer before he was elected to the Academy. A chemist, E. V. Murphree, likewise published but 26 papers. Other physicists and chemists wrote as many as 380 papers in this sample, and half a dozen scientists in yet other fields wrote less than 50. Thus, differences among sciences are not so important as those among scientists and their publishing habits.

In sum, it is by no means necessary to write an enormous number of papers to be a successful scientist, and yet a quarter of our small sample wrote more than 250 — 1 every two months for 40 years. Did this serve any useful purpose other than to keep the scientist from being underfoot at home or from engaging in some socially undesirable activity? For that matter, is writing so many papers a socially desirable or undesirable activity? Unquestionably the papers overload journals, which is undesirable, but to examine these questions in a significant way we must turn from papers to citations, which give a measure of sorts of the contribution papers make to science.

[2] National Academy of Sciences, 1877 and since, *Biographical memoirs* (New York, Columbia University Press). Data are from Memoirs 9-16, 29, 39, 40, with individuals selected to obtain a range of fields and periods of publication and, in part, for ease of identification in comparing with the Institute for Scientific Information, 1968, *Science Citation Index* (Philadelphia, Institute for Scientific Information).

TABLE 5.2. Publication Records of Some Members of the National Academy of Sciences in Different Fields

Name	Publishing Dates	Field	Number of Papers
Abbot, Larcom Henry	1857–1909	Civil Engineering	140
Bailey, Solon I.	1893–1931	Astronomy	95
Barrell, Joseph	1899–1927	Geology	90
Boltwood, Bertram Borden	1895–1923	Chemistry	38
Boss, Lewis	1879–1919	Astronomy	130
Brewer, William Henry	1851–1904	Agriculture	130
Bumstead, Henry Andrews	1902–1921	Physics	18
Chamberlin, T. C.	1872–1928	Geology	251
Chandler, C. F.	1856–1873	Public Health, Industrial Chemistry	38
Clark, William Bullock	1887–1917	Geology	110
Clarke, Frank W.	1868–1928	Geochemistry	400
Clarke, John Mason	1877–1913	Paleontology	440
Cope, Edward Drinker	1859–1915	Paleontology	1,395
Coulter, John Merle	1881–1929	Botany	240
Crafts, James Mason	1862–1915	Physical Chemistry	140
Dana, James Dwight	1835–1895	Geology	220
Fewkes, Jesse Walker	1874–1928	Archeology	270
Forbes, Stephen Alfred	1870–1930	Entomology	400
Gooch, Frank Austin	1888–1918	Chemistry	130
Hague, Arnold	1870–1913	Geology	45
Hall, Granville Stanley	1866–1924	Psychology	439
Hilgard, Eugene Woldemar	1854–1916	Agricultural Chemistry	320
Hunt, T. Sterry	1846–1891	Geology	360
Loeb, Jacques	1884–1926	Physiology	405
Mall, Franklin Paine	1887–1921	Anatomy	104
Minot, Charles Sedgwick	1868–1913	Embryology	190
Nichols, Ernst Fox	1894–1925	Physics	26
Osborne, Thomas B.	1884–1929	Biochemistry	252
Packard, Alpheus Spring	1860–1914	Zoology	579
Pickering, Edward C.	1865–1918	Astronomy	266
Prudden, Theophil M.	1879–1924	Pathology	82
Remsen, Ira	1869–1923	Chemistry	90
Ridgway, Robert	1869–1929	Ornithology	540
Rosa, Edward B.	1889–1922	Physics	90
Sargent, Charles Sprague	1874–1927	Botany	110
Smith, Sidney Irving	1864–1890	Marine Biology	70
Trowbridge, John	1871–1911	Physics	80
Verrill, Addison Emery	1861–1926	Marine Biology	293
Wells, Horace Lemuel	1885–1923	Chemistry	75
Wilczynski, Ernst J.	1895–1923	Mathematics	77
Wright, Arthur Williams	1870–1911	Physics	32

Source: National Academy of Sciences, 1877 and since, *Biographical memoirs* (New York, Columbia University Press). Data from Memoirs 9, 12, 13, 14, 15, 16.

Papers and Citations and Scientific Fame 95

TABLE 5.3. Productivity of Papers by Some Members of the National Academy of Sciences Compared with Citations in the *Science Citation Index* for 1968

Name	Field	Publishing Dates	Papers	Papers Cited	% Cited	Total Citations	Maximum Citations/ Paper	Biographies in Last 10 papers	Dates of Cited Papers
Allen, Charles Elmer	Genetics	1901–1945	46	7	15	10	2	1	1903–1935
Atkinson, George F.	Zoology	1886–1919	180	4	2	4	1	1	1895–1909
Bucher, Walter H.	Geology	1911–1963	100	15	15	23	4	1	1919–1963
Clausen, Roy Elwood	Genetics	1912–1957	44	9	20	9	1	0	1944–1957
Coblentz, Wm. Weber	Physics	1903–1953	380	22	6	23	2	2	1905–1949
Dodge, Raymond	Physiology	1896–1941	86	10	12	26	6	0	1900–1926
Dryden, Hugh L.	Engineering	1920–1966	220	33	15	56	9	0	1933–1965
Eisenhart, Luther P.	Math	1901–1963	140	50	36	81	8	0	1909–1966*
Goldschmidt, Richard B.	Genetics	1903–1960	270	21	8	28	3	1	1908–1955
Hoagland, Dennis R.	Biochemistry	1913–1951	85	43	51	98	19	0	1923–1950
Ives, Herbert E.	Physics	1906–1951	236	19	8	32	5	0	1908–1939
Keith, Arthur	Geology	1891–1935	41	4	10	4	1	2	1905–1913
Kidder, Alfred V.	Archeology	1901–1961	180	4	2	4	1	3	1932–1937
Lamb, Alfred B.	Chemistry	1902–1945	45	14	31	17	2	0	1911–1945
Landsteiner, Karl	Medicine	1892–1943	290	75	26	148	9	0	1899–1963*
Millikan, Clark B.	Engineering	1929–1963	41	3	7	4	2	2	1938–1939
Moore, Joseph H.	Astronomy	1903–1952	148	12	8	16	5	2	1909–1948
Murphree, Eger V.	Chemistry	1923–1949	26	2	8	2	1	0	1925–1943
Ritt, Joseph F.	Math	1913–1951	71	5	7	6	2	0	1922–1950
Ross, Frank E.	Astronomy	1905–1943	70	8	11	12	4	0	1920–1935
Seares, Frederick H.	Astronomy	1896–1952	140	4	3	6	3	2	1922–1928
Seashore, Carl E.	Psychology	1893–1950	238	6	3	9	3	1	1908–1939
Szilard, Leo	Physics	1925–1964	30	16	53	32	9	0	1929–1964
Walcott, Charles D.	Paleontology	1875–1931	280	4	1	4	1	2	1878–1914
Woodworth, Robert S.	Psychology	1897–1959	200	49	25	114	41	5	1899–1965*
Wright, Frederick E.	Geophysics	1898–1953	192	8	4	8	1	2	1908–1934

Source: National Academy of Sciences, 1877 and since, *Biographical memoirs* (New York, Columbia University Press), data from Memoirs 9–16, 29, 39, 40; Institute for Scientific Information, 1968, *Science Citation Index* (Philadelphia, Institute for Scientific Information).
* So says the *Citation index*.

CITATIONS

It is commonly recognized that prestige in science attaches to those who already have it. For example, reprints are sometimes filed not under the name of the unknown first author but under that of the famous second or third author, because that way they can be relocated. Merton has documented this very well and calls it the "Matthew effect" from a quotation from the Gospel according to St. Matthew: "For unto every one that hath shall be given, and he shall have abundance; but from him that hath not shall be taken away even that which he hath."[3]

An aspect of this effect can be quantified most readily by merely taking all or part of the literature of a given year and determining the distribution of citations among scientists. This yields the general result that among those scientists who are cited at all, the top 10 percent receive 50 percent of the citations, the least cited 50 percent receive only 10 percent, and the intermediate 40 percent receive 40 percent of the citations.[4] However, studies of this sort beg the more important question of what fraction of scientists or scientific papers are cited at all and what is the citation distribution among all papers. Price has suggested that "on the average, every scientific paper ever published is cited about once a year."[5] He derives this from the average rate of literature expansion (7 percent) and the average number of citations (15) in a paper. This yields 105 citations to each 100 papers existing at that time.

The actual distribution of citations can be determined if the population of papers as well as of citations is known. The annual rate of citation can be found by a careful comparison of the *Science Citation Index* with the papers published by certain individuals. Temporal variations in the rate of citation are much more difficult

---

[3] R. K. Merton, 1968, "The Matthew effect in science," *Science 159*, 58.

[4] See, for example, Derek J. de Solla Price, 1965, "Networks of scientific papers," *Science 149*, 510–515, for a study based on data in the *Science Citation Index* computerized files for the single year 1961. If the citation index data are accepted as published, they tend to underemphasize the occurrence of multiple citations, because the computerizing retains erroneous citations. Thus, initials are transposed or one initial used instead of two or names misspelled, which may turn one real scientist who is cited many times into a small group who are little cited. Likewise, many people with identical initials and last names have lived in the last century and a half, and they became one scientific giant to the computer.

[5] Ibid., p. 511.

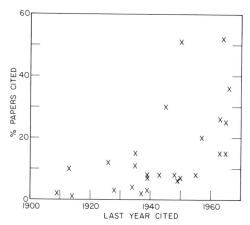

Fig. 5.2 The decline with age in the fraction of papers by NAS members that are cited in 1968. All papers by a member are plotted at the date the last was published.

to measure, because they require historical citation indexing and thus reference to a flood of original literature. Both are attempted here.

It is reasonable to assume that the members of the National Academy of Sciences are among the top 10 percent of scientists who receive 50 percent of all citations, and thus it is of interest to determine the citation distribution to their work (Table 5.3) in the *Science Citation Index* for 1968. Data on publications and fields of study are from the *Biographical Memoirs* of the Academy. The 26 scientists wrote about 3,780 papers, of which 447 were cited for a total of 776 citations. As we expected, all the Academy members were cited, but unexpectedly only 12 percent of their papers were. How can this be compatible with Price's average rate of citation of one per paper per year? There is some correlation between frequency of citation and time of research. The fraction of papers cited for each person decreases from about 30 percent for those last publishing in 1965 to 3–5 percent for those last publishing in 1915 (Fig. 5.2). We have found by a comparison of growth rates that citations to work in a given field are in a relatively constant proportion regardless of the date of publication. This seems to suggest that few scientists are now engaged in investigating the same problems that interested Academy members early in the century.

An alternative possibility is of fundamental importance with re-

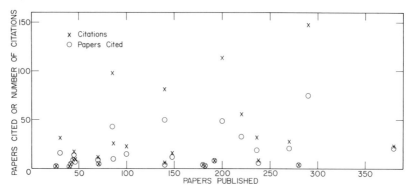

Fig. 5.3 Number of cited papers and citations to them compared with number published by NAS members.

gard to the outpouring of scientific literature. The tables in this chapter show that the number of papers per highly productive scientist has gradually decreased. Thus, although the percentage of older papers cited is smaller than that of modern ones, the number per author has changed much less if at all. The number of cited papers per year per person seems to approach an expectable maximum of 20 for a range of published papers from 30 to 380. Only a fifth of the sample group exceed this number (Fig. 5.3). This suggests that the run-of-the-mill distinguished scientist may not be capable of writing more than a rather limited number of distinguished papers.

One way to pursue this point is to test it with a totally different population in a historical study of citations. This has been done with a group of geologists who seem reasonably typical in that among those cited, the top 10 percent received 65 percent of the citations, and the bottom 50 percent received 8 percent.[6] The population studied consists of 3,078 authors who wrote 16,224 papers, or an average of about 5 each. Only 669 people were cited, but they received an average of about ten each, or 6,646 in all.

Unlike most studies of citations, this one can assess the proportion of scientists who are uncited as well as those who are. This gives a very different picture of the magnitude of the Matthew effect.

---

[6] The population is that of people with names beginning with "A" or "B" in the *Bibliography of North American Geology* from 1785 to 1960. All of these names were checked for citations in each paper in each volume of the *Bulletin of the Geological Society of America* from 1888 to 1969. Thus, the discussion pertains not to all citations but only to all in the largest and most influential journal for the last 80 years. Warren Smith and Joel Ryan did most of the compilation.

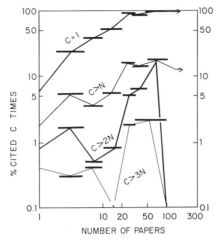

Fig. 5.4 The fraction of authors writing N papers who are cited C times in the literature of geology.

Fully 78 percent of all geologists receive no citations; 11 percent receive 8 percent of the citations; another 9 percent receive 27 percent; and to the top 2 percent of the population are given the abundance of 65 percent of all citations.

We shall turn to citations to individuals, but it is revealing to consider the probability that a scientist who writes n papers will receive c citations. This is shown in Figure 5.4. We may put the results in terms of the various stages in the career of a research scientist. In the first stage he writes a paper. In the second stage he has one cited. The chance that he will reach this stage depends on the number of papers he writes; for one paper it is 6 percent, for 10 papers it is 50 percent, for 20 papers it is 90 percent, and it is virtually certain for more than 50 papers. Of course, this may not happen immediately. One author of 24 papers was cited for the first and only time 49 years after reaching stage one. Stage three is when $c > n$, the scientist has more citations than papers. This does not mean all papers are cited. The chance of reaching stage three is 1.7 percent for the author of one paper, about 5 percent for 2–19 papers, and 15 percent for 20–100 papers, above which the chance decreases. The fourth stage is when citations are more than double papers and, as the graph shows, it takes about 20–30 more papers to have the same chance of reaching this stage as the one below. That is, it takes at least 20 papers to have a 5-percent chance and 50 to have a 15-percent chance. The highest stage identified for citations in a single journal is three times as many citations as papers. The chance of reaching this is about 2 percent for authors of 20–100 papers, but very rare for the less-known authors of fewer papers or for the famous ones who write more than the literature can cite. In sum, a multistage career is not beyond the grasp of the author of a single notable paper, and his chances improve markedly if he writes more, but only up to a certain limit. In this we find another indication that a scientist may benefit little or may actually suffer from pouring out papers.

In these citations we have yet another means of testing whether there is a negative correlation between productivity and recognition or utility of a scientist's papers. We have already found that the probability of being cited more than n times decreases beyond n equals 100 papers. What of the actual number, c, of citations — does it continue to increase even though the ratio of c/n decreases? The answer is probably not. The percentage of people cited 50 to 100 times is about the same for those who write 50–100 papers as for those who write more. Moreover, with very few exceptions there appears to be a limit of about 35 papers cited regardless of the number written. The number cited doubtless would increase if all journals were included in the study instead of merely the principal one. Nevertheless, we have a confirmation that most good scientists are capable of producing only a small amount of good science. They can either distribute it in a few or many papers, depending on their style and urge to write, which appear to be independent characteristics.

We can gain another insight into this question by studying the probability that a first paper will be cited. All the cited papers by authors of one paper are first papers, and we have found the probability is 6 percent. This rises to 12–15 percent for authors of 2–19 papers and is 25–30 percent for those writing 20–100 papers (Fig. 5.5). For those who write more than 100 papers the chance that the first will be cited drops to 9 percent, which appears to be significantly less. The last sample is only 11 people, and it may be meaningless. However, Kirk Bryan, the only one who had a first paper cited, barely qualifies for this group; moreover, within the scope of this study he never had his second through seventh papers cited and his first one only 35 years after it was published. The writer of a single paper cannot cite his own work, but this is not true of more prolific scien-

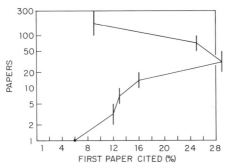

Fig. 5.5 The fraction of authors writing N papers who have their first paper cited in the *Bulletin of the Geological Society of America*.

tists who, in fact, are more apt to cite themselves than is anyone else. Thus, if we are attempting to evaluate a scientist's contribution or prestige, we must subtract some proportion of his citations as being self-generated. Consequently, except for the 6 percent, all the probabilities should be somewhat reduced. After this correction it appears that the first papers written by the most prolific authors have little more chance of being cited than those of authors with a single maiden effort. Considering the visibility or professional fame of prolific authors, this seems surprising. Presumably the scientific content of papers is closely related to the probability of citation. If so, the content in papers by writers of minimal and maximal output tends to be much less than for those who lie between the extremes. I suggest that this means a scientist is part puzzle-solver and part author. The minimal publisher is exhausted in one or both capacities. With regard to geologists, those publishing less than 20 papers may be stronger in writing than in puzzle-solving, considering the many criteria by which their output seems substandard. The author of 20–100 papers has both abilities strongly developed and tends to have a maximum amount of puzzle-solving per paper. The highly prolific authors of more than 100 papers may be even stronger in both abilities, but are more apt to be normal in puzzle-solving and outstanding only in authorship.

The actual number of papers in these stages may vary from science to science and may be lower in most than in geology. Nevertheless, it appears that prolific authors tend to overdo it whether they are distinguished or not. The necessity for publishing or perishing, and for writing papers in order to be paid to attend meetings, and the invitations to write just one more review, are detrimental to the quality of science in many ways. Evidently almost all scientists and science itself would benefit if the paper blizzard were properly viewed as pollution rather than desirable output. Is there any reason why most scientists should not be limited to 2 or 3 papers in any given year and a total of 15–20 per decade? It would make for more contemplation and would not affect promotion if the limit were generally applied. It would minimize repeated publication of the same paper under different guises. People might actually read more papers instead of merely abstracting them and KWIC indexing them for computers.

Administration poses only trivial difficulties. Abstract journals are already computerized and issued monthly. A simple integration of the tapes would yield a list of the ten least-wanted paper polluters, and their papers would thenceforth be rejected for a year or a decade. The problems would be with the early careers of young scientists and the later careers of authentic scientific giants. Perhaps young scientists should be allowed to publish at will for the first decade. By that time positions, promotions, and recognition should be reasonably settled. Then the limits on production could go into effect.

The superstars remain. In 1968 alone, Einstein had 252 citations to 118 papers and 21 to one paper. Linus Pauling had more than 1,000 citations, of which roughly 150 were to the *Nature of the chemical bond* published in 1960. Chandrasekhar also had more than 1,000 citations to 300 papers and 65 to a book published in 1961. Scientists of this caliber should receive every encouragement to publish, but how are they to be separated from the common herd? The career stages based on the ratio of citations to papers should serve for this purpose. At the end of a novitiate decade, a scientist goes under the limit system and stays there until he reaches stage three, when the ratio is greater than one. That should help him concentrate on quality. Those who are at stage three can produce without limit unless they drop into a lower stage. At the end of a second decade, a superstar might be expected to maintain stage four and so on, like the Canadian Air Force Exercise schedule. I suspect that the suggested career-stage levels are set much too low and the annual and decade limits are too high, but they could be carefully adjusted at appropriate intervals. It is evident that citations are far from a perfect measure of contributions to science, and we shall analyze some of the other factors that influence them. However, they seem to be the best thing available, and proper computer programming could take other factors into account.

Whether this system goes into effect or not, it seems worthwhile to venture some generalizations about which papers are apt to be cited. For the most part there are very few citations to the first five papers written by even the best scientists. Presumably this is because they are just getting the hang of research, and most are engaged in solving trivial problems which are then presented in obscure journals. Moreover, they tend to produce abstracts, which are hardly ever cited no matter when or by whom they are published. It is not obvious how this can be remedied. Possibly scientists, like many other authors, should simply burn their initial works. This seems unlikely because, unlike playwrights for example, they are rewarded whether the work is very good or not. Perhaps this observation will merely serve to limit discouragement when initial papers seem to be ignored.

Although there will have been no hint of it to this point, it also is true that there are very few citations to the last five papers written by even the best scientists. In part this may be because men eventually run down, but an important factor is that many of the papers are biographies and obituaries of colleagues. It is possible that these are cited by biographers and historians who, however, rarely write for scientific journals. It is readily apparent that the sixth paper is a critical one in a scientist's career. If it is an obituary notice, then he probably should abandon any hope of being cited and reaching stage two.[7] Will his writing have been for naught? Not at all. Beyond the possibility that he may have done something for science, he surely has done something for himself. He writes papers and therefore is promoted over the man at the next desk who does not. He writes obituaries and so grossly improves the probability that someone will write one for him — which is, after all, more than most men leave behind.

REGINALD ALDWORTH DALY: A MAN FOR ALL SEASONS

We have examined some statistical results of citation analysis, but some kinds of information can only be derived from a more detailed look at citations to an individual. For this purpose I have selected

---

[7] Not always. National Academy member Henry Bumstead wrote a biographical memoir as his second paper.

R. A. Daly, who published first in 1896 and last in 1952. For most of that time he was Sturgis Hooper Professor of Geology at Harvard, where I first saw him in 1947 — a tall, straight man with a cane which he waved while giving characteristically detailed, accurate, and vigorous directions to a little Chinese boy who apparently wanted to find a building in the Yard. Daly was at the very least among the leading geologists of his time and received practically every honor in the science. He wrote on a wide range of subjects and presented great integrating hypotheses in most.

He conceived major theories about the origin of igneous rocks, the interior of the earth, the effects of changes of sea level during ice ages, the effect of gases in vulcanism, and the origin of submarine canyons. For forty years a significant fraction of the significant geological literature consisted of tests and problem-solving related to Daly's ideas. In many papers on the origin of igneous rocks in the 1920's and 1930's the authors did not formally cite him, but merely referred to "Daly's hypothesis" or "Professor Daly's ideas" with the quite valid assumption that any reader knew about them. It is a pleasure to study an aspect of the career of such a man.

The data consist of citations to Daly's papers in several major journals from the time he first began to be cited through 1968. There are separate annual tabulations of citations to each paper and of the number of citations by each person who cited Daly. For long periods the citations are identified by journal. With these data we are able to trace the citation history, and thus the contribution, of individual papers and see whether annual citations decay in the manner indicated by tabulating the ages of citations in a single year. We can also see whether citations and citers follow distribution functions like so many aspects of science and society.[8]

[8] The procedure was to record citations in the *Bulletin of the Geological Society of America* from 1890 through 1968, in the *Journal of Geology* from 1905 through 1968, and in the *American Journal of Science* from 1921 to 1968. The first of these was used as a control, and the other starting dates were selected by tracing backward from 1925 to a period when he was uncited for several years. Citations in these older journals are very trying to count because of the style of reference. After about 1950 it is quite easy, as all references are tabulated at the end of a paper. The technique used after initial familiarization was to examine the table of contents in a journal, see which papers were on subjects on which Daly had written, and examine their bibliographies. All doubtful papers were examined and occasionally a whole journal, so as not to miss some unsuspected style of citation. In general, however, the citation process is very regular for long periods. The same people write in the same journals on the same subjects and cite the same papers. The study captured at least 90 percent and probably far more of the citations in the three journals. A note for students: always

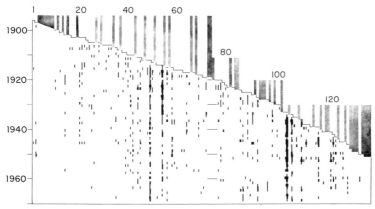

Fig. 5.6 The distribution of citations to the publications of R. A. Daly in three major journals of the earth sciences. The titles of the more cited ones are: no. 49; *Geology of the North American Cordillera at the forty-ninth parallel*, 857 pp.; no. 54, *Igneous rocks and their origin*, 563 pp.; no. 103, *Igneous rocks and the depths of the earth*, 598 pp.

Daly wrote 136 papers, of which 86 received 859 citations. We immediately note that fully a third of the papers by even the greatest geologist were never cited. Like lesser men, he also went for a long time with hardly any encouraging citations. His third through ninth papers were never cited and his first not until 19 years after publication (Fig. 5.6). His first citation came on his second paper in 1898, and then eight barren years passed until suddenly 5 different papers were cited. Our study is capable of identifying when the subject cites himself, which we might expect to be disproportionately frequent in his early career. Indeed, 4 of these 6 citations were self-generated.

By 1911, Daly had written 45 papers and received 23 citations. None of the papers would receive more than 15 citations in our study in the next 50 years; only a quarter received more than 5. Daly passed stage one in his career in 1896 and stage two in 1898, but if he had continued on in the same style he probably never would have reached stage three, when citations exceed papers. In 1912 and 1914, however, he published two books which immediately began to collect several citations per year each and eventually accounted for 78 and 82 cita-

---

work forward in time except for the very beginning. After a while it is very easy to identify papers which should cite a scientist. Thus, it is a fascinating and time-consuming puzzle when he is not cited. By working forward you avoid the further complication of wondering why a paper was not cited when in fact it had not yet been written.

tions respectively. The first of these was *Geology of the North American Cordillera at the forty-ninth parallel,* a memoir of the Canadian geological survey 857 pages long. The second was *Igneous rocks and their origin,* a book of 563 pages. As a consequence of writing these books, Daly reached stage three in 1917, when he had 65 papers and 77 citations, and stage four in 1926, with 91 papers and 186 citations. Until 1932, he did not write another paper which ever totaled 20 citations.

In 1933, he published his most successful citation-getter, *Igneous rocks and the depths of the earth,* which netted 126. It was a book extending the ideas in his earlier one on igneous rocks, and its publication cut citations to the first one to less than a quarter. This important work was shortly followed by *The changing world of the ice age,* another book which collected 43 citations, and by *Origin of submarine canyons,* a paper which got 22. From 1933 to 1936 was the flowering of Daly's career; he had been publishing for 40 years. By 1937, his citations tripled his papers, but he never again published a heavily cited paper. Even his book *Strength and structure of the earth* in 1940 received only 21 citations.

The distribution function for citations to his papers is that the top 10 percent of the publications received 52 percent of the citations, and the bottom half received only 9 percent. We have already found that only 2 percent of earth scientists receive 65 percent of all citations. Daly is certainly in the highly cited elite. If his work is typical, a mere $10\% \times 2\% = 0.2\%$ of publications collect about a third of all citations.

Daly's most cited works are all books; does this mean that there is a constant ratio of citations to pages, and thus it makes little difference how the pages are packaged? We know that this cannot be correct for scientists as a whole from the data already discussed, but it might more plausibly apply to the work of a single author. For the books, the ratio ranges from 5 to 20 pages of publication to catch a citation sometime. This is a mere fourfold ratio, which suggests a far more linear relationship than for most distribution functions. For shorter papers, however, the range is from 1 to 30 pages to catch a citation, and in this we can see the variations in quality in which all scientists faithfully believe. Counting the uncited papers, the range is 1 to infinity, which gives a more than adequate basis for faith.

We turn now from the citations to the citers. We identify 334

Papers and Citations and Scientific Fame 107

citers and the range of citations per person from 1 to 79. The maximum number was of course by Daly himself, which means that, at the time he stopped publishing, he had made 12 percent of his citations and in 1968 it was still 9 percent. The next most frequent citer was Frank Grout with 28, followed by Washington with 22, and Flint and Shepard with 16 each. It is noteworthy of Daly's range of interest that two petrologists, a geomorphologist, and a marine geologist are so grouped together. The top 10 percent of the citers gave 45 percent of the citations. For the whole distribution we have the following: the author cites 10 percent, the other members of the top group cite one-third, the middle 40 percent cite one-third, and the bottom 50 percent cite one-fifth. This is a much more even distribution function than for authors versus citations or authors versus publications. By sending copies of his reprints to only 32 colleagues, Daly could have made access to his work easy for those who would give him half his citations. Presumably these were the major, prolific contributors in his subfields. With a total of 60 reprints he could have rendered the same efficient service to those who would cite him more than 3 times and thus probably wrote more than one paper in his subfields. We cannot but wonder what happened to and what was the function of distribution of hundreds of reprints, which was not an uncommon practice before the dawn of xerography.

What of the 194 people who cited Daly only once? They range from the minimal producer in his subfields to the prolific man in another subfield or science whose interests minimally overlap Daly's. The common characteristic of the whole group is that they are less likely to be familiar with the literature in Daly's subfields than are those who specialize in the subfields. This is the group who are most apt to be influenced in citations by Daly's general scientific prestige and visibility. Perhaps the unexpectedly large proportion of single citers reflects an increment attributable to this effect.

What happened to the intensity of citation to individual papers through the years? Normally, data on the decreasing use of citations come from studies of ages of citations in a single volume of some journal. This does not say what happens to an individual paper, as can be done with the present data. The sample involves little more than three journals which grow only 40 percent in volume through 50 years. It takes no account of the vast increase in number of journals that gives the doubling of literature. Thus, if the citation half-

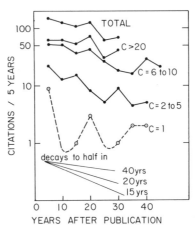

Fig. 5.7   The number of citations, C, received by papers of R. A. Daly in 5-year intervals after publication, plotted in groups according to C.

life is correctly analyzed, the chance of a citation to a given paper should decrease to one-half in about 15 years. This is confirmed for all of Daly's papers in Figure 5.7 that give a decay to one-half in 15–20 years. However, it appears that the history of citations to a paper varies with the number of citations that it receives. Data are inadequate to analyze this effect for papers cited only once, but they suggest that the decay rate is relatively rapid. For papers cited 2–5 times, the half-life is about 20 years or about the same as average for the whole group. For papers cited more than 20 times, the half-life seems to be about 40 years, and it is intermediate for papers with intermediate citations. If the literature expansion is taken into account, this means that the most cited papers actually are cited more frequently with increasing age. Apparently, papers cited only once or twice tend to be those of interest only briefly after publication. Those cited many times in general are far ahead of their time, and thus are cited with research-front intensity until the impacted front finally leaves them behind.

SLOWLY GROWING FIELDS

The production of papers and the patterns of citations they catch are quite regular for normally expanding science. How are they in fields and subfields with abnormal growth rates? We shall first consider the patterns in slowly growing or steady-state subfields, using data from a historical analysis of American geologists and their works.[9]

To begin with, the distribution function for the productivity of scientists which is such a typical characteristic in all normal fields is markedly different in steady-state subfields. In chemistry and physics and in geology during the nineteenth century (all of which expanded normally), the number of scientists who write n papers decreases at $1/n^2$ from 1 to about 15 papers and then about $1/n^3$ for larger numbers of papers. The relationship is quite different for such steady-state subfields as invertebrate paleontology and geomorphology. In these the number of scientists drops off much more slowly than $1/n^2$ for about 20 papers. The proportion of people in the subfields who write 10 papers is about four times as great as in chemistry and physics. For 20 papers the proportion is almost seven times greater. From 20 to about 80 papers the number of invertebrate paleontologists and geomorphologists who write n papers appears to drop off as $1/n^2$ and for larger numbers it varies as $1/n^3$. Thus, for the most prolific people in these subfields, the curve is just like chemistry and physics.

These relationships can also be expressed in terms of the cumulative number of people who have written *at least* n papers instead of the number who write exactly n. For ordinary physicists and chemists who write less than 10 papers, the cumulative number decreases as $1/n$, and for more prolific writers this distribution function varies as $1/n^2$ (Fig. 5.8). This distribution function for geomorphologists and invertebrate paleontologists varies as $1/n$ from 20 to 80 papers and as $1/n^2$ beyond. Thus, for comparison in terms of appointments and promotions, it appears that a specialist in a slow subfield, at least in geology, is required to write 20 papers to qualify as an ordinarily productive scientist and more than 80 to be considered prolific.

If 20 is the minimum number fitting the ordinary distribution function, what is the significance of a distribution function of roughly $n^{1/2}$ for those who write less in a slow subfield? In terms of achievement, these people seem to stand in the same relationship below ordinary scientists as distinguished ones do above. If, as Price sug-

[9] This analysis is based on the 3,078 people indexed in the various issues of *Bibliography of North American Geology* whose names begin with "A" and "B." The number of papers per person is the total for the period 1785 through 1959. Thus, still active people will ultimately have larger totals. The specialists in various subfields are identified by the titles of their papers.

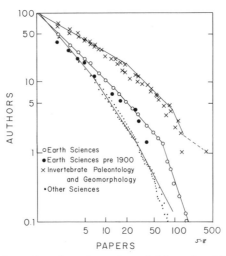

Fig. 5.8 The number of authors in various fields and subfields who write at least N papers plotted as a function of N and normalized to a maximum of 100. The points for chemistry and physics are after Derek J. de Solla Price, 1963, *Little science, big science* (New York, Columbia University Press), p. 18.

gests, ordinary scientists double in 10 years and distinguished ones in 20, this group should double in 5.[10] The slower subfields appear to be plagued by a rapidly growing number of minimal-level scientists or technicians who write papers. This hardly augurs well for American science if a general steady-state or arithmetical growth begins. Most science is produced by a small fraction of distinguished men. Price has noted that in normal science it is necessary to quadruple the ordinary men in order to double the distinguished ones. We now find that in slow or retarded science the number of literate technicians must increase sixteenfold in order to double the men at the top. The promotion and prestige criteria are such that papers will be authored by ever-lower levels of technicians. Presumably most of the papers will be of a type an ordinary scientist could write with his mind functioning at the square root of optimum and a distinguished scientist if he is merely awake. Most, as we shall soon see, are immediately and forever ignored in research, but they fill journals and bibliographies and abstract journals and computerized data systems and contribute to the strangulation of science.

Modern geology, as a whole, is a little abnormal, but is not in such

---

[10] Price, *Little science*, p. 39.

a parlous state. From 1 to 20 papers the distribution function is a little less than $1/n$, and from 20 to 70 papers it is $1/n$; for larger numbers it is $1/n^2$. Thus, as a whole, it differs from other sciences chiefly in that distinctively prolific people are required to write more than 70 papers instead of more than 10. The geology curve in Figure 5.8 is intermediate in position between other sciences and the slow subfields of geology. This shows that some other subfields have distribution functions essentially like chemistry and physics. Presumably these are the rapidly growing ones such as oceanography and isotope geochemistry. Since almost everyone in these subfields is alive and active, the subfields cannot be analyzed in the same way. However, geology as a whole was mostly young and vigorous before 1900. For that period the distribution curve for 1 to 10 papers is the same as for chemistry and physics, but it deviates for larger numbers. Thus, the creation of a class of author-technicians occurred largely in the steady-state period.

Citation distributions also are affected in steady-state science. We can draw on the data for individuals in steady-state subfields within the general study already used in this chapter to analyze citations in geology.[11]

Because the number of pages is relatively constant, the number of citations should decay to half in 20 years or more if it is linked to the slow growth rate of the subfields. A sample of about 500 citations shows a decay rate which more closely approximates 15 years for 30 years or two doubling periods (Fig. 5.9). All subdivisions of the sample more or less follow the same distribution pattern except papers with only one citation, which decay more rapidly. The single paper with more than 50 citations does not provide an adequate sample, but the trend suggests increased citation as time passes. This again suggests that papers with many citations improve with age compared to those with few.

The data are not clear about whether citation decay and literature growth are closely related. We have found that in some slow subfields

---

[11] The geologists selected are the eleven with more than 50 papers or 50 citations who published in geomorphology or paleontology other than micropaleontology during the period 1920 through 1939. This gives at least 30 years of data with minimal biasing because of the citation blizzard late in the dormant period. Older papers cannot be included in a coherent sample because of the tripling of pages in the *Bulletin of the Geological Society of America* in the early 1930's. The number of pages varies hardly at all in the remainder of the period analyzed.

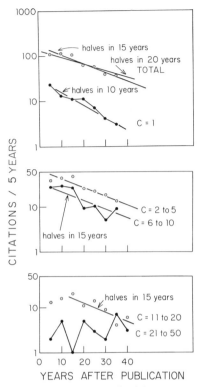

Fig. 5.9 The number of citations, C, received by papers of various prolific authors specializing in invertebrate paleontology and geomorphology in 5-year intervals after publication, plotted in groups according to C.

there is a disproportionately large fraction of minor authors; they are rarely cited more than once or twice. Even the prolific authors in slow subfields have citation distributions quite unlike Daly's. I have lumped citations to eleven prolific geomorphologists and paleontologists distributed through a long time range. The citation distribution is that the top 10 percent of papers receive only 37 percent of citations, and the bottom 50 percent receive 20 percent. This composite result may be compared with citation analyses for two outstanding geomorphologists, Eliot Blackwelder and Kirk Bryan. Respectively they had 44 percent and 36 percent of their citations for the top 10 percent of papers, and 17 percent and 18 percent of citations for the bottom 50 percent of papers. Papers with few citations appear to have a shorter half-life than ones with many. An abundance of

Papers and Citations and Scientific Fame

such papers might account for a citation decay rate greater than the growth rate for literature in the subfield.

The data consistently indicate a period of more intense citation from 5 to 15 years after most papers are published. This may be the effect of the citation blizzard in the dormant period. Perhaps it may indicate that citations in the first 5 or 10 years after publication are inadequately represented compared to livelier subfields. With a slower pace of publication, it may take a decade for the research front to reform to the point where new work builds on old and cites it.

CITATIONS IN A RAPIDLY GROWING SUBFIELD

I hope the reader will excuse me if I report the result of an analysis of citations to my own work as the example in a fast subfield, namely, marine geology. Much of the labor in these lengthy analyses is minimized if the subject is familiar at the beginning. Some avenues of investigation were suggested in the study of Daly's citations which could only be followed by questioning the author at great length. These are open to study in what follows.[12]

Publications of which I am first author number 45, began in 1948, and were subject to analysis through 1968, by which time 225 people had given 676 citations to 40 papers. The time span might seem too short for a significant comparison with the 75 years in the study of Daly's work. However, it covers four doubling periods and thus a sixteenfold expansion of the literature in the subject and of the specialists concerned with marine geology. This is an expansion comparable to that in the subfields of Daly's principal interest during 75 years.

I need hardly say that to the best of my knowledge the first publication has never been cited. However, it was in vertebrate paleon-

---

[12] With Daly I sought to deal with the chief journals over a long time span. This is impossible in a rapidly growing field, because it would go through so many doublings it would fragment. Consequently, I studied all thirteen journals in which I was cited at all frequently. In this analysis, therefore, the effects of literature expansion are included. The journals include *Nature* and *Science* and five journals that began publishing in 1960 or later. Marine geology grew so rapidly that it saturated the journals, and *Marine Geology* was founded in 1961. Even that could not contain the supply, and numerous papers appeared in books which were little more than bound volumes of proceedings of special meetings. At times in the early 1960's papers in the subfield were as common in such books as in regular journals, and they are included in this study.

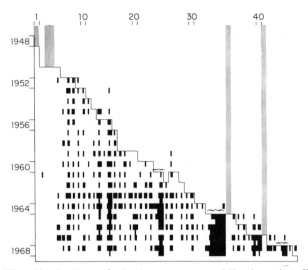

Fig. 5.10 The distribution of citations to my publications in all principal journals in the earth sciences plus multiauthored books. Titles of the more cited publications are: no. 15, "Deformation of the Northeastern Pacific and the west coast of North America," no. 24, "The East Pacific Rise," and no. 33, *Marine geology of the Pacific,* a book of 271 pages.

tology, which is among the slowest of subfields. Only one of the first four papers received a citation, and that was after 13 years; meanwhile the first citations were after only 4 years, and then there were suddenly a total of 6 citations to 4 different papers. Five of the 6 citations were to my first papers about marine geology. All were cited within a year of publication, and this is the first indication of a far more intense citing than in slower subfields. Several of these papers have since been cited more than once per year (Fig. 5.10), and consequently only a year later the citations exceeded the number of papers, and by 14 years after the first citation there were ten times as many citations as papers.

The same levels of intensity compared to slowly growing fields are shown by papers on the "Deformation of the Northeastern Pacific Basin and the west coast of North America" and on "The East Pacific Rise," which received an average of 5 and 10 citations per year respectively. The other major collector of citations is a book, *Marine geology of the Pacific,* which was cited about 30 times a year. Another indication of growth and intensity of a subfield is the frequency of citation in the year after publication. None of Daly's

Papers and Citations and Scientific Fame

papers was cited more than once in the first year, nor twice even in the second year according to our analysis. Very few papers in slow subfields were cited at all in the first 2 years. We know that Daly and the others were cited very heavily for people in normal or slow geological subfields, which makes the contrast with marine geology particularly strong. *Marine geology of the Pacific* was cited 24 times in the first year and several short papers were cited 11 or 12 times. Indeed it is not uncommon in the subfield for several citations to appear only a few months after publication. This is made possible by exchange of preprints. The post and the telephone echo with the plea, "Make up your mind what you are going to call that paper so I can cite it."

The distribution function for citations to these papers in a fast subfield is very like those in normal science.[13] The chief difference is in the proportion of papers with no citations at all. If this is taken into account, the distribution functions are even more similar. I suppose that this means that after a sufficient number of doubling periods the chance that one paper will be cited more than another one by the same author depends mainly on his style of writing. The chance that the aggregate of his work will be cited frequently depends to a great extent on growth rates, but the division of the work among shorter and longer publications is what tends to determine this distribution function.

The distribution function for number of citations per citer is also very similar in the two studies — none of the proportions differing by more than 3 percent. Again it appears that the function is related not so much to growth rates, which might affect when the citations are made, but to the basic structure of the process of citation. In this case the matter is subject to further analysis. The citers can be divided into four classes according to whether they are (1) contemporaries, that is, they began publishing within a doubling period of 5 years before or after me, or (2) older, or (3) younger by one to three doubling periods, or (4) even younger — this last group are now graduate students or gained doctorates within the last few years. Of the 225 people who have cited my work, I find that I know: all who have made more than 4 citations; a decreasing fraction of these citing 4 to 2 times, but even so, no less than 80 percent; and

---

[13] The top 10 percent of the papers received 47 percent of citations, and the bottom 50 percent received 8 percent.

slightly more than half who have cited only once. This is by casual inspection of the names and with no further effort to refresh a faulty memory. It is immediately clear that the citations are largely from what Price calls an "invisible college" consisting of a working group interested in the same research problems. However, it is by no means an entirely contemporary group.

The most frequent citer other than myself (88) is Bruce Heezen (36), who is an exact contemporary often doing much the same thing in a different ocean. Seven other people have made 10 or more citations and 3 more are contemporaries; 1 is slightly older, 2 are younger, and 1 is much younger. Of this group I note that with some overlap, 4 have been my coauthors and 4 have been shipmates on long expeditions. From this small, almost contemporary group come almost a quarter of all the citations. The people making 6 to 8 citations are oddly distributed among the groups. None are contemporary, 5 are older, 7 are younger, and none are much younger. The last group is easy to explain; few will have had the time to make so many citations. The younger group, on the other hand, are people working in exactly the same specialty and with much the same techniques as I, but who have not been at it quite so long as some of my contemporaries. The citations by the older group are mostly by prolific scientists with many fields of interest, of which perhaps a quarter overlap mine. I find I have been at sea with just half of the colleagues who have cited my work more than 5 times. The gap in citations with a frequency of 6–8 contemporaries is puzzling, but may merely be a statistical oddity. If it is verified by other studies, it would suggest that within the invisible college is a highly visible central tower surrounded by a moat.

We turn to the effects of growth rates upon citation distribution. Data are plotted in the same manner as for citations to Daly, except that the time scale is expanded to accommodate the brief doubling time (Fig. 5.11). The only papers that can be used are those more than 10 years old which give 2 doublings. As a whole, they indicate not a constant rate of citation but rather an increasing one. As observed with Daly's citations, it appears that the distribution function varies with the number of citations to a paper. Data are too sparse for consideration of papers with less than 6 citations. For the 6 to 10 citation papers, the highly irregular curve may be fluctuating around a

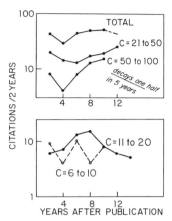

Fig. 5.11 The number of citations, C, received by my papers largely in marine geology in 2-year intervals after publication plotted in groups according to C.

horizontal axis. For the next larger group, the trend is rising slightly, and for 21–50 papers it rises by 25 percent in 10 years. The effect is even more striking for the only paper with more than 50 citations. Its citations double in 10 years.

These results indicate that the more-cited papers increase in proportion to the less-cited ones, regardless of the growth rate of the subfield. The Matthew effect is again confirmed.

The number of citations in papers in rapidly expanding fields also seems abnormal. Plate tectonics is an exploding subject in geology, and the first paper appeared in December 1967. Papers typically and naturally have been published in minimum-delay journals such as *Nature,* and the half-life of citations is only a few years. Minimum referencing might be expected, but a sample of 16 papers gives an average of 34 citations each, with 3 with over 50 citations and also 3 with less than 20.[14] This striking characteristic appears also in the *Journal of Molecular Biology,* in which 13 papers contain an average of 27 citations each, and 6 have more than 30.[15] These papers have the other characteristics of those in plate tectonics, namely, rapid publication, explosive growth, and very brief half-life of citations. Papers in *Physical Review Letters* also share these characteristics, but they contain an average of only 10 citations each; thus, not all rapidly

[14] From the author's reprint collection.
[15] *Journal of Molecular Biology 39,* 158–243 (1969).

expanding fields have a superabundance of references.[16] In the *Journal of the Optical Society of America*, the odd circumstance occurs that the papers with few citations tend to cite older work as would be expected in a field which is expanding only slowly.[17] In sharp contrast, the half-life of citations is only 5 years in papers with more than 30 citations. Thus, the "reviews" tend to be in rapidly expanding subfields. In sum, it is a rather common characteristic of the average research paper in a rapidly expanding field to contain the number of citations more commonly associated with review papers.[18] This seemingly odd property probably arises because of the same need that causes review papers to be written. A review paper is needed not because there are many papers to be consolidated, but because there are many potential readers who do not know what is in the many papers. In normal science, and particularly in steady-state science, this circumstance generally arises because a new crop of scientists have not studied all the old and very old literature, or because it has been widely diffused in obscure journals. In an exploding field the same need arises, but for a different reason. Much of the communication is verbal, by meeting abstract, by preprint, by review of proposals, and the like. The publications cited commonly have appeared so recently that reprints have not yet been distributed. An author has a very real reason to believe that most of the readers will not have had time to become familiar with the work he is citing.

The bountiful citation practice in rapidly expanding fields presumably influenced the number of citations to my own work in marine geology. What of the increase in citations as time passes? All papers plotted have 6 or more citations because the sample is too small to consider for papers with less. Likewise, it is difficult to determine the distribution function for rarely cited papers by other marine geologists, because such a large fraction of the papers are too new to be indexed and the authors too young to be identified. Thus, we have no direct evidence of what happens to rarely cited papers. On average, however, the expanding citation of some papers must be balanced by unusually rapid decay of citations to others, and these are the ones with few to begin with. If the speed with

---

[16] Derek J. de Solla Price, 1969, "Citation measures of hard science, soft science, technology and non-science," paper presented at Conference on Communication among Scientists and Technologists, Baltimore, The Johns Hopkins University.

[17] *Journal of the Optical Society of America* 60 (1970), Jan. and Feb.

[18] See Chapter 6.

which the Matthew effect prevails in normal science is normalized to the sedate pace of a roulette wheel, then in fast subfields it is akin to a crap game.

We can now reexamine the interpretation that the observed balance between the expansion of the literature and the half-life of citations indicates a broad generalization that on average each paper ever written is cited once per year. We find that most papers are never cited or are rarely cited and have short half-lives. Others are cited frequently and have long half-lives. Thus, even as a first approximation, it appears misleading to think of the average paper as being cited once a year. It is more realistic, but still a generalization, to say that although the number of citations to the publications of a given year remains constant, the fraction of the papers that are cited is small to begin with and decreases with time. The Matthew effect is again operating. The age distribution of citations calculated from studying the literature of a single year is a highly insensitive indicator of the distribution of citations to individual papers. Most papers are uncited; an increasing fraction vanish into oblivion; and fewer and fewer papers slowly receive more and more citations per year.

CITATIONS IN A SCIENTIFIC REVOLUTION

The earth sciences are in a revolution because of the demonstration a few years ago that continents drift, the sea floor spreads, and the earth has had a lively history. A few details of the demonstration are needed in order to understand the significance of a citation analysis of the subject. Harry Hammond Hess published what he called an "essay in geopoetry" in 1962. He suggested that the mantle rises under midocean ridges in the center of the ocean basins and is converted by a reversible reaction to oceanic crust. It then spreads laterally, rafting continents with it, to oceanic trenches where it is reconverted to mantle. His reasons were multitudinous, but included the facts that the ocean basins contain no old rocks, the sediment is too thin for the basins to be old, various criteria suggest mantle convection under the oceanic crust, and so on. In the following year Fred Vine and Drummond Matthews took note of the fact that the earth's magnetic field reverses itself episodically and that the orientation of the field is recorded in the magnetic minerals in cooling lava flows. Thus, they reasoned that the cooling and spreading of lavas in the center of ocean basins should produce magnetic stripes parallel to the central ridge. They accepted Hess's hypothesis, but did not cite his 1962 paper.

The stripes were soon found, and by overwhelming evidence they were shown to be caused by the suggested reversals of the magnetic field. The stripes are bilaterally symmetrical for thousands of kilometers on each side of a midocean ridge — which has been explained only by Hess's spreading mechanism. The stripes have been dated, and the distance from the center is proportional to the age. Thus, it is even known that the spreading has had a constant speed for a long time. Most geological evidence is relatively vague because of the many uncontrolled variables involved. It is important to emphasize that this evidence is quite adequate to convince mathematicians and physicists. These measurements plus countless confirmations showed that the fundamental elements of Hess's argument are correct and gave the foundation for the present scientific revolution.

By tracing the history of citations to Hess's paper, we can examine how well they reflect the impact of a truly important contribution. Regrettably the picture is clouded. Hess originally expected to publish his geopoetry in volume III of a multiauthor work called *The sea: ideas and observations.* He cannot have felt any urgency in publication, because it is well known that such volumes are invariably delayed for unconscionable periods while the most indolent authors try to meet their commitments. He did the standard thing, issued a mimeographed preprint, which most of us received in 1960. No one got very excited, but confusion resulted when Robert Dietz published very similar concepts in *Nature* in 1961. Hess then, or meanwhile, gave up *The sea* and published his paper, "History of ocean basins," in yet another symposium volume called *Petrologic studies: a volume to honor A. F. Buddington,* published by the Geological Society of America in 1962. Dietz tidied things up in 1963 by giving priority to Hess for the basic ideas of interest here. Considering the prevalence of multiple discovery, this was a most unusual act. Hess then created some bibliographic complications by writing a further elaboration of his hypothesis in yet another symposium volume in the *Colston papers* in 1965. Thus, in 1960–1962, anyone wishing to refer to these ideas could cite either Dietz or the preprint by Hess. Then for two years either of two papers might be cited, and after

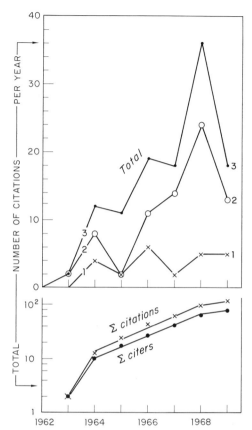

Fig. 5.12 Citations to a revolutionary paper, "History of ocean basins," by Harry Hess.

1965 any one of three. Nonetheless, the 1962 paper by Hess was by far the most frequently referenced.

Citations to Hess's paper increased fairly regularly from 2 in 1963 to 36 in 1968 and then declined to 18 in 1969.[19] This gave an exponential doubling in about 17 months from 1964 to 1968 and then a marked retardation of growth (Fig. 5.12). The books and journals studied can be grouped according to the publication delay in order to see if the citing authors had any sense of urgency about priorities.

---

[19] We have examined the following journals from 1962 through 1969: *American Journal of Science, Journal of Geology, Marine Geology, Bulletin of the Geological Society of America, Science, Nature, Journal of Geophysical Research, Earth and Planetary Science Letters, Tectonophysics,* and the *Journal of Petrology*. We have also studied citations in seventeen symposium volumes resulting from meetings during the period.

Apparently they were concerned, because in every year but 1965 most of the citations appeared in journals with the fastest available publication. In 1965, three symposium volumes swamped the citations to the Hess paper.

We can identify 76 authors who cited "History of ocean basins" in the literature studied. Their number increased almost as rapidly as the citations. The distribution is 58 authors who cited once; 12 twice; 3 three times; 2 four times; and 1 seven times. In only eight years, few of them had much time to cite the paper more than once. Even so, the prevalence of single citers suggests some factor at work. To understand it we must first look at what the authors said about the paper when they cited it.

As we have already observed, one of the principal reasons for the dormancy of the earth sciences was a surfeit of geopoetry without the means to distinguish the work of a Keats from that of an Eddie Guest. All the facts that Hess explained were already explained by other hypotheses. Consequently, the real novelty in Hess's paper was that he *called* it geopoetry and that he explained so many different things. Eleven citations were given in 1963 and 1964. One was by Hess who said "if it is correct" and one by Wilson who had very similar ideas anyway. These 2, plus those by Orowan and Schuiling, who are applied physicists, gave a total of 4 citations that accepted the idea of sea-floor spreading. The 7 citations by geologists and geophysicists responded to the idea with a hardly disguised "Hohum." One paper merely cited some data from Hess and never referred to his ideas at all.

A paper by Maurice Ewing and others in 1964 set the tone for many in 1965 and even 1966. Ewing and his colleagues stated that the object of their study of sediment distribution and thickness in the North Atlantic was to test the hypotheses of continental drift and of sea-floor spreading.[20] If Hess was right, they reasoned, the sediment would be thin in the center of the basin where the crust is newly created, and it would thicken on the increasingly older crust away from the center. What they found is that the sediment is indeed thin or absent in a central band 75 miles wide, but that in the surrounding region it is uniformly thick. This was baffling and remains so. Consequently, in the summary of conclusions, they say

20. M. Ewing, J. Ewing, and M. Talwani, 1964, "Sediment distibution in the oceans: the Mid-Atlantic Ridge," *Geological Society of America Bulletin* 75, 17–36.

nothing at all about the avowed object of the study except that "the sediment distribution also denies the possibility that any part of the ridge crust, except the crest, is appreciably older than another" (p. 33).

Given apparently conflicting data, attention was focused on similarities rather than differences. The time was not ripe for a revolution.

Precisely the same thing occurred in a paper by van Andel and colleagues published in 1965.[21] In a comparatively detailed geological survey, they found that by every criterion available to them the crest of the Mid-Atlantic Ridge is young and the flanks are older. However, the conclusion they reached was: "We do not, however, believe that our data either require or materially support such an interpretation" of sea-floor spreading (p. 1216). Their conclusion was absolutely correct because the data were subject to other interpretations and the time was not ripe.

I myself went through this phase at about the same time. After offering a hypothesis to explain many aspects of marine geology, I pointed out that the "bolder hypothesis" by Hess would also explain them, but that the facts did not require acceptance of his ideas. Dale Krause in 1965 was in about the same position when he found his data "in harmony" with sea-floor spreading but also with other ideas. By the end of 1965, the count was 12 citations by unconcerned or skeptical authors, 8 citations by those who had faith, and finally a crucial citation by Vine and Wilson who gave proof of sea-floor spreading.[22] Proof does not mean instantaneous enlightenment for all scientists. The overwhelming power of the proof was only gradually apparent to most of us.

In 1966 and 1967 the members of the marine geology establishment looked at their data and went through the same experience. Most said, "My observations are not compatible with sea-floor spreading, and I shall prepare a critical demonstration that this is so and thus demolish this nutty idea and we can all get back to work." One by one they found, for they were honest scientists, that in fact their data, regardless of the subject, were compatible with sea-floor spread-

[21] T. H. van Andel et al., 1965, "Morphology and sediments of a portion of the Mid-Atlantic Ridge," *Science 148*, 1214–1216.
[22] F. J. Vine and J. Tuzo Wilson, 1965, "Magnetic anomalies over a young oceanic ridge off Vancouver Island," *Science 150*, 485–489.

ing. An elaborate network of confirmations appeared, and most marine geologists got back to work but with a new paradigm. During this period several people reexamined data on which they had already published interpretations not favorable to Hess's hypothesis. The time was ripe, and gradually the various uncertainties in interpretation were found to balance in favor of the hypothesis. Several ad hoc explanations were still available for most facts and relationships. What was changed was the paradigm, and it because of a few crucial observations and field experiments. The citers in 1968 and 1969 were not the established marine geologists who were on to other things by then. Mostly they were very young marine geologists and the members of the establishment of continental geologists who were examining *their* data in a new light. Almost all accepted the proof of sea-floor spreading and began to renterpret what they had already done.

We can now return to the question of why so many authors cite "History of ocean basins" only once. A scientific revolution means that a large number of apparently unrelated facts are seen to be in an unsuspected relationship. Many established workers have available the means to apply critical tests. Each type of test is important, but there is little occasion for many repetitions. Thus, one author, one subject, one test, one citation.[23]

The graph (Fig. 5.12) suggests that the 36 citations of 1968 may not again be matched. The number probably will drop off to some much lower steady-state value. It appears that the extremely important papers that trigger a revolution may not receive a proportionately large number of citations. The normal procedures of referencing are not used for folklore. A real scientific revolution like any other revolution is news. The *Origin of species* sold out as fast as it could be printed and was denounced from the pulpit almost immediately. Sea-floor spreading has been explained, perhaps not well, in leading newspapers, magazines, books, and most recently in a color motion-picture. When your elementary school children talk about something at dinner, you rarely continue to cite it.

---

[23] I have cited the paper seven times. The fact came as a surprise to me and certainly was not deliberate. As it happens we had at Scripps: (1) some brilliant students, (2) an enormous stock of unprocessed magnetic data, and (3) a data-processing capacity which my associates and I had developed to analyze echograms. The instruments and techniques could also analyze the magnetic data just when it was demonstrated that they provide the best information about sea-floor spreading. There was a windfall.

We have attempted to verify some of the citation patterns of a revolutionary paper by a brief study of the very famous one in which Watson and Crick announced the structure of DNA.[24] Citations in the *Journal of Molecular Biology* have remained relatively constant from 1959, when the journal was founded, to the present. The only major perturbation was in 1964–1965, not long after the authors and M. H. F. Wilkins won the Nobel Prize for this work. In the *Science Citation Index* the annual number of citations from 1964 through 1968 was 48, 49, 45, 40, and 36 respectively, and it is estimated at about 30 for 1969.[25] This suggests a declining trend, but also a paper that will be cited extensively for some time to come. To a geologist, the number of journals in which citations occur is astonishing, and it would be interesting to follow the expanding recognition of the importance of the paper and its broad applicability. By 1965, it was being cited in the *Journal of Dairy Science* and the *Journal of the American Oil Chemists Society,* among others.

CITATIONS, PUBLICATIONS, AND ACHIEVEMENT

Publications and citations are related in ways that depend on rates of growth. The absolute size of a science, subfield, or specialty determines how many citations are subsequently granted to the publications of a given year. However, the history of citations to a paper depends not on what went before or what happens when it is published, but only on what happens afterward. It is growth, not size, that determines the incidence of citations. Most papers are uncited, which means, disregarding quality, that no more papers are written on the same detailed specialty. This subject has a zero growth rate. In a subfield with a normal growth rate, papers steadily accumulate, and they necessarily cite those that came before, but with a most uneven distribution. Even in a normal subfield, many papers are dead ends in narrow subjects with a zero growth. They are, by definition, uncited. Some papers are in relatively hot subjects, and they are cited relatively quickly and frequently. Inasmuch as

[24] J. D. Watson and F. H. C. Crick, 1953, "Molecular structure of nucleic acids — a structure for deoxyribose nucleic acid," *Nature 171,* 737.

[25] This paper gives a neat example of the difficulties of using these very useful volumes. Sometimes the journal is listed as *Nature* and sometimes it is separately listed as *Nature London.* We have identified the cited year as 1953, 1958, and 1963; the volume number as 171 and 177; the page number as 737, 373, 727, 137, and 780.

they tend to be in the mainstream of science, they may become even more frequently cited as the hot subject expands.

A subfield contains hot specialties, and a science has hot subfields with rapid growth rates. In these the citation phenomena are about the same as in normal subfields, but they are more intense. Regardless of the size of such a subfield, its papers on average are more frequently cited than those in a normal subfield. We visualize two subfields, a large one, P–1,000, with normal growth and a small one, P–10, with fast growth. P–1,000 can generate about 3,000 papers per year and thus a maximum of 3,000 citations to a single paper. P–10 can give only 30 citations. However, 3,000 citations occur only if all the new papers are about the same subject as the single paper. If so, it is a subject, by definition, with extremely fast growth, and the other subjects in the subfield are neglected. The average rate of citation for the whole subfield is unaffected and remains so unless the whole growth rate is changed. The members of P–10 are more cited than those of P–1000.

What, then, makes a subject grow and causes a scientist to be cited? The invention of a totally new instrument, such as a microscope or laser, may open a whole new world to study and is followed by a flood of citations to the inventor. This does not occur steadily in any science, and it has not been a common occurrence in geology. In fact, one of the main reasons for the dormant period was that very few new instruments and techniques were invented.

Citations in the environmental sciences are commonly given by field workers to pioneers who preceded them into a region. Likewise, the discoverer of some field phenomenon is cited if it is found elsewhere. These two factors put a great premium in geology, as in war, on being there "fustest." Daly was cited repeatedly for his report on the exploration of the 49th parallel, and Bryan for his work on the Papago country in the southwest. This is one reason why the sea and the moon, the only places available to explore, are so attractive for young scientists.

Our study of citations to authors with names beginning with "A" and "B" reveals another way to garner many citations, namely, by influencing students. Doubtless, individual professors and university departments have the best interests of students at heart, and so it is only natural that Professor P doing research in subject X should encourage his students to do the same. Often they select him for a

Papers and Citations and Scientific Fame

doctoral thesis advisor, because of his ability to project his enthusiasm for X. Students receive degrees and publish their theses and usually continue on with the same research for some time. Perhaps they have students of X in their turn. All the resulting papers ought to cite Professor P who, by simple tabulation, is thus a famous man and receives an endowed chair. This leads more students to his side. The problem is that the ability to enthuse students may not be closely related to the ability to identify and pursue important research problems. A great scientist may inspire the choice students of a major university, but so may a literate technician inspire the marginal students at Backwash U. Both will be frequently cited because they automatically generate a rapid expansion of the subject. Consider the abundant literature on the size-frequency distribution of sedimentary rocks. It appeared at one time that this distribution might reveal much about the origin of the rocks. The needed equipment was inexpensive, the techniques were simple, and hundreds of theses and publications appeared describing such distributions. Each had the usual list of citations to earlier literature. In retrospect nothing much came of all this effort, and one wonders if it would have continued so long had it been difficult and expensive. If not, the numerous citations would not have existed. This would not have affected the reputations of the pioneers, because it was worth trying and they pioneered other things either before or after. For the followers, however, it would have been another matter.

A little counting of citations can be a dangerous thing. The quality or even the real impact of a scientist's work should not be measured out by something simply because it is the only thing which can be measured. Citations probably are related to the importance of a paper, but this is certain only if "importance" is defined as having the same subject as some later papers. Citations measure a rate of reproduction and do not distinguish the breeding of Darwins from that of Jukeses and Kallikaks. The undesirable breeders of science are supposed to be sterilized by the review process before publication, but it catches only the eccentric not the establishment dullard. A limit on the permissible annual output of scientists would at least act as a birth control for everyone.

On the other hand, every successful scientist is a breeder of successors in research, and it is reasonable to assume that those who breed many citers have more impact than those who do not. Thus, citation indexing may provide a valid means for comparing people working in the same subfield or on the same subject. The best man working on X can be identified even though citations don't tell us how he compares with the best, or even the worst, specialist on Y. For this a better measure might well be the I.Q. or Graduate Record Examination scores of the students of X and Y.

6 Scientific Literature

The literature of science is viewed by some as a means of communicating scientific results and by others as a social device for determining priority of discovery and thereby establishing prestige. However, both of these purposes apply equally well to the scholarly literature of the humanities. Does scientific literature differ from non-scientific only in its subject matter? It has been argued that the particular characteristic of science and of its literature is that the paradigms are almost universally accepted at a given time. Thus, a firm foundation is available on which to build a new edifice rapidly and efficiently.[1] It is at least arguable that an essential characteristic of scientific literature is that it grows rapidly.

We may visualize a group of scientists confronted with a new problem and with the tools to solve it. They make the necessary experiments and publish the results. This continues until the methodology no longer yields significant new results. The standard procedure then is to use the same tools to attempt to solve a different problem or else to develop new tools to work on an untractable facet of the old one. In either event, the literature grows rapidly. Consider, on the other hand, the poor wretches in a government laboratory directed to work with the original equipment on a problem they have long since solved. Will they publish vigorously, and if so, about what? What happens in a science in which neither the paradigms nor the technology are capable of dealing with the most fundamental problems of interest? We have seen that in geology

[1] Derek J. de Solla Price, 1969, "Citation measures of hard science, soft science, technology and non-science," paper presented at Conference on Communication among Scientists and Technologists, Baltimore, The Johns Hopkins University.

the literature grew only very slowly. We may now consider whether its content remained normal science — in which case growth rate does not affect it — or whether it began to bear some resemblance to the literature of nonscientific fields.

Let us establish a model of the development of the scholarly literature of a completely sterile subject in order to see what we should look for in the literature of dormant geology. Consider the number of angels dancing on the head of a pin. A normal scientific literature includes measurements of the area of pin heads, the size of angels based on visual sightings, and calculations of packing factors for different dances. This results in order of magnitude estimates. New graduates entering the field publish theses consisting of ever more refined measurements. Senior scholars see to the establishment of Project Skywatch and document its progress as a project. At this point a sensible scientist goes on to some other subject.

The scholars remaining in the field and their students focus their attention on such subjects as "The meaning of the word *angel*." Other groups classify all known types of pins and dances, and the discovery of each additional type results in another paper. Different schemes of classification require prolonged and heated publication, because each is arbitrary and most are equally valid. There is increasing concern with the new terminology of classification, and jargon flourishes. This is accompanied by ever more stifling and capricious editorial restrictions. A technicalese is substituted for English in the name of clarity. The promulgators of this technical jargon are noted for their cry, "Scientists cannot write English."

Ultimately the literature of the field becomes its subject. Scholarly discourse is no longer expected to lead to a conclusion. It is sufficient to list all proposed explanations for the size of angels without attempting to test them. Bibliographies abound. Ancient papers are discovered by ever more intense delving in the stacks. The most ancient papers are cited more frequently than when they were new. Naturally all this scholarship is time-consuming. The literary polishing, checking of jargon, and redoing of bibliographies when the editors once again change the rules also take their toll. Publication is at a much slower rate than in the initial stages of the investigation of angels and pins. Compare, for example, the attitude that it takes less time to go make a new measurement in the ocean than it does

Scientific Literature

to try to find an unindexed old one somewhere in the international data bank.

CONCERN WITH STYLE

What happens to the output of papers in a stagnating field? We may conceive that a spiritual malaise generated by lack of solvable problems will discourage some scientists. However, this is unlikely to suppress a field so successfully as happened in geology in the first half of the century. Writing scientific papers is like robbing banks. By the time you get good at it, you tend to be cut off from other activities and it is awfully hard to stop. The productive scientist who doubts this should try kicking the habit for a year — even six months. Thus, if the output in geology was stifled, we should not attempt to ascribe it to any decline in energy or dedication on the part of authors. Instead, in simple justice, we should look for some counterforce of equally energetic and dedicated men seeking to prevent publication. Administrators can stop scientists from doing significant work by limiting support, but they cannot prevent publication of trivia. Thus, the only likely candidates for a counterforce are the editors and critics who enter directly into the publication process. If there are a hundred writers to each editor and critic, we may expect the publication pipeline to flow. If the ratio is reversed, the flow will stop. Presumably some intermediate ratio existed in American geology from 1920 to 1955, with the result that the viscosity of flow greatly increased and the velocity slowed.

We can seek to document this presumption by looking at the output of the Geological Survey, which was the component that slowed the most. In the nineteenth century, survey authors were highly prolific. In 1909, the survey issued *Suggestions to authors* as a style guide for this great outpouring. The second edition came out in 1913, and the third, which ultimately went to seven printings, appeared in 1916. Clearly the survey contained a significant cadre of incipient editors and critics. Is it entirely a coincidence that the output of survey publications almost ceased by 1920? The fourth and, one hopes, last edition of *Suggestions to authors* appeared in 1935. We may assume that a generation of survey geologists spent an entire career surrounded by new copies and editions of this work.

Twenty years, and perhaps the demise of a generation, were to elapse before the output of survey publications once more became vigorous.

Could a mere style guide really have such an effect? Not by itself, but it could if everyone in the survey was trying to follow or even emulate the guide. Consider the opening sentence:

> Although the geologist may with advantage include in his field notes carefully written descriptions that can be transcribed literally into the manuscript of his report, he should generally not attempt to dictate offhand from his notebook with the intention of rearranging and polishing the typewritten matter thus obtained to form a final report but should study and classify his notes and material before he begins to write.

The cadence seems similar to that of Bryant's "Thanatopsis," and the subject is about as lively.

The examples of editing are highly instructive. Unfortunately, the guides are full of introductory sentences replete with the same literary flaws which they then proceed to dissect in writings of others: for example, "Adverbs and adverbial phrases are by some writers commonly misplaced . . ."

The survey author had to follow an elaborate style guide, and this doubtless slowed his output. However, that put him over only the first hurdle in a very long race. Page 3 in the fourth edition of *Suggestions to authors* states:

> All Survey manuscripts are prepared for the printers by the editorial staff in the section of texts. The editorial work includes the examination of the character and gradation of headings, the form of footnotes, the use of geographic and geologic names, the form of tables and sections, and the various features of typographic style — such as sizes and styles of type, capitalization, punctuation, and spelling — as well as many other details.
>
> Much of this work is done according to prescribed rules, such as those of the *Style Manual* of the Government Printing Office, or mandatory decisions, such as those of the United States Geographic Board, the Division of Geographic Names, or the Survey's committee on geologic names.
>
> The editorial work includes also suggestions to the author concerning the arrangement of matter, paragraphing, the correction of faults or errors in grammar or rhetoric, the clarification of obscure passages, the elimination of repetitious or irrelevant matter, and many other features. . . .

All of this editorial work takes people and time, and publications are delayed and fewer. The bulletins themselves are admirable sources with which to document delays carried to excess. They give not only the date of a publication, but also a completion date of the field work on which it is based. We may randomly consult bulletins 1031 A through 1031 E which were based on field work completed from 1951 to 1956 and published from 1955 to 1962. The maximum delay was eleven years, which merited a note: "This report in essentially its present form and under the same title was released to open file by press notice No. 90568 dated December 6, 1955." [2] In short, the survey authors had become reduced to writing nonpublications in order to overcome the delays in the system.

It might appear that these delays were some late manifestations of a congestion of manuscripts left over from wartime. This conjecture was tested by examining a bulletin from just before the war. *Bulletin* 920, published in 1941, contains a poignant statement which proves that the delays were long standing.

> Most of the field work upon which this report is based was done about 1906, in anticipation of the publication of a folio of the Geological Atlas on the area. The long delay in publication . . . [p. 4]. In the interval of 25 years between the first and last visits to the area many of the old mines and quarries had been abandoned, their dumps had been removed, and it became impossible to discover their former sites. Nevertheless, the manuscript has been left essentially as it was originally written . . . [p. 5]

This was the United States Geological Survey in its hour of travail. The distinguished geologists of the survey readily circulated from government to university and back. They were not only officers in professional societies but also reviewers and critics and even editors of journals issued by those societies. The end result was that a sizable and influential fraction of American geologists became accustomed to prolonged delays in publication and extensive editing of style, and developed a deep concern for the literature as such. It was only natural that they concerned themselves ever more frequently with publications about old literature rather than about new scientific results.

[2] J. C. Miller, 1962, "Geology of the waterpower sites in Alaska," *Bulletin of the U.S. Geological Survey 1031E*, p. 104.

BIBLIOGRAPHIES

Without bibliographies a great deal of time would be lost in random searches of the literature, and thus they are valuable. To the mature specialist in an active field they may not be of much interest, because most of the literature has been written during his career and he is familiar with it. Consequently, he tends to remember that Smith wrote concerning the alkali rocks of Fiji in about 1962, and when it is of interest he consults his reprint file to see what Smith said. The situation is quite different in dormant or slowly growing or even very old fields. No one is alive who read most of the literature as it accumulated. Both mature scholars and students have an acute need for bibliographies to help sift through a mountain of paper. This suggests that students should avoid fields with extensive bibliographies, and specialists should become uneasy when the first lengthy one appears. A bibliography identifies not only a distressing amount of material with which one must be familiar, but also hints at the existence of a labor surplus in the field. This does not apply if the compiler is himself a professional bibliographer. If so, a change in the object of investigation is implied rather than unemployment.

I am about to draw upon the various volumes of the *Bibliography of North American Geology,* which is the best testimony I can give for their value to me. I seek to examine the growth of specialized bibliographies and the intensity of the bibliographic component of geology as a function of time. Some bibliographies are essential to the advancement of science. However, if all the literature is bibliographies, then there is no science even if the bibliographers are scientists. Where is the dividing line?

The early bibliographies identified in the *Bibliography of North American Geology* are all concerned with publications of an individual geologist of the sort normally presented in an obituary biography. By 1873, however, papers on the geology of Minnesota were sufficiently numerous to be the subject of a bibliographic compilation. Two bibliographies of invertebrate fossils were issued before 1880, and thirteen more before 1890. This illustrates the fact that the second bibliography is easy to build on the first and more and more inevitably follow. Is there any other explanation for the publication of the second bibliography of the geology of Minnesota be-

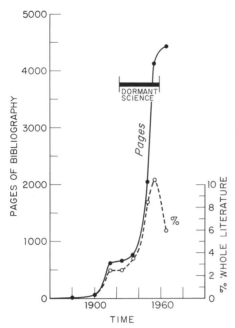

Fig. 6.1 The emphasis on formal bibliography during the dormant period in the earth sciences. The list of references in scientific papers constitute informal bibliographies, and these varied in the same way.

fore the first appeared on any other state? Bibliographies continue to accumulate and reproduce. During the 1940's the Geological Society of America published in its *Memoirs* and *Special papers* some 1,779 pages of bibliography of fossil invertebrates. During the 1960's the society has already published 3,367 pages of bibliography on fossil vertebrates compiled by a distinguished group of specialists whose hopes that this work will be definitive may, I fear, be disappointed.

Let us attempt to see if there is any correlation between the preparation of bibliographies and the dormancy and reawakening of geology. We can compare the pages of bibliography published in a year with the total number of pages of geology, which we already know.

I have selected convenient single years in each decade for the analysis, summarized in Figure 6.1. In 1860, there were no pages of bibliography; in 1880, there were two pages or 0.03 percent of the whole literature. This increased to 59 pages or 0.3 percent in 1900, when the literature was still growing rapidly. By 1914, the growth had begun to stall and the pages of bibliography increased rapidly to 620, which was fully 2.5 percent of the whole. This percentage remained constant in 1925 and increased to only 3.5 percent or 760 pages in 1935. By 1948, the science of geology had been in the doldrums for about three decades or much of a professional lifetime. Few people were active who had seen the bulk of the literature grow around them. The men needed bibliographies, and they got them. The number of pages increased to 2,050 and the percentage to 8.5. This trend continued to 1954 when the numbers were 4,140 pages and 10.4 percent. More than a tenth of the whole output consisted of literature about literature at the end of the dormant period. The reawakening of the earth sciences began shortly; the most vital growth occurred in new or small fields lacking bibliographies.

The increased and logical use of modern data-processing made it relatively easy to produce 4,420 pages of bibliography in 1965, but this represented only 6.0 percent of the rapidly expanding literature.

It is distressing to contemplate, but it must be realized that the amount of "informal" bibliography produced during the dormant period was also significant. In the next section I shall show that in 1940 and 1950 roughly half of the geological literature had the characteristics of review articles with regard to number of citations. The first three issues of the *Bulletin of the Geological Society of America* for 1940 contain 488 pages of articles, of which 38 are bibliography and 17 are editorial matter. If this was typical of the period, the total number of pages of bibliography accompanying journal articles may have been equal to that in formal bibliographies. Conjecturally, then, the percentages of pages given above might be only half of the total.

How this will all end is uncertain. Perhaps the percentage of bibliography will continue to decline if the literature and the science continue to grow in new directions. However, it is clear that in a dormant science increasing emphasis was placed on the literature of literature. For American science in general, if fiscal winter comes, can bibliographies be far behind?

CITATIONS

The use of citations at different times gives us an additional criterion for possible overconcern with literature. Price has shown that scientific papers cite an average of 15 older papers, but that these are unevenly distributed.³ Research reports constitute 75 percent of the papers, and of these 40 percent contain 0–10 citations, and the remainder have 11–25 citations. At the other extreme are review papers, which consist of two groups: reviews constituting 5 percent of the whole literature and with more than 45 citations, and about 1 percent of superreviews with more than 84 citations.

We may compare the geological literature with these norms by counting citations in the first 10 to 20 papers in appropriate issues of the *Bulletin of the Geological Society of America*. The results are given in Table 6.1. It is often suggested that environmental sciences

TABLE 6.1. Distribution of Citations in Papers in Earth Sciences at Different Times Compared with an Average Distribution for Science as a Whole

		Number of Citations (Percent)					
		Less than 10	11–25	26–44	45–83	More than 84	Total Number	Average per Paper
Average for science		49%	35%	10%	5%	1%		
Earth sciences	1880	60	33	7	0	0	196	15
	1910	30	50	20	0	0	170	17
	1930	42	42	8	0	8	226	19
	1940	17	17	25	25	16	947 (436)	79 (40)
	1950	47	13	27	13	0	376	25
	1960	38	34	14	14	0	446	21
	1965	25	35	25	15	0	525	26

Source: *Bulletin of the Geological Society of America* for the years indicated.

like geology and ecology have characteristics different from physics and chemistry. This table suggests, once again, that the relationship is misunderstood. Geology is unlike the physical sciences when it is growing slowly, but it is similar when it grows rapidly. In 1880 and in 1910, the average number of papers was normal and so was the distribution. If anything, references were relatively few — suggesting

³ Derek J. de Solla Price, 1965, "Networks of scientific papers," *Science 149*, 510–515.

a predominance of research papers over reviews. The situation was not very different as late as 1930, although the average number of citations increased. By 1940, geology had been dormant for decades, and the pattern of citations was grossly changed. One paper in this bulletin was in fact a bibliography of 512 references to fossil man; if it is included as part of a random sample, the average number of citations is 79. Even if it is excluded, the average number is 40 and the distribution is quite abnormal. The research paper was largely replaced by the review article, or, in other words, most papers were bibliographies. It appears that this condition continues. The proportion of review articles and the average number of citations both remain high. There was some respite around 1950 when "Notes" began to appear in the *Bulletin*. By 1960, these amounted to 40 percent of the papers published in the first three months. They typically contained about six references, were short, and were published promptly. In brief, they were normal research papers. Moreover, they tended to have different subjects from the remaining papers in the *Bulletin*. In 1960, most were concerned with the new and rapidly growing field of radioactive dating of rocks. The proportion of notes was similar in 1965, but the publishing delay even for "Notes" was often a year or more. These data do not identify any pattern associated with the reawakening of the earth sciences in recent years. The publication delay times may indicate why. The normal research style of citations returned to geology when new fields of geophysics, geochemistry, and marine geology, among others, began to predominate over the old. Specialists in these new fields turned to other journals which published faster. This is observation, not speculation.

Some insight into the concern with literature when a science is dormant is given by the decay time for citations. In normal science, if the references cited in a journal volume are counted by date of publication, their number decreases to $1/2$ in 15 years.[4] Let us see how the geological literature compares with this average value. We use the same volumes and papers which were just examined with regard to the number of citations. In 1890, the half-life was far less than 10 years, and none of the literature was more than 35 years

[4] Derek J. de Solla Price, 1963, *Little science, big science* (New York, Columbia University Press), p. 79.

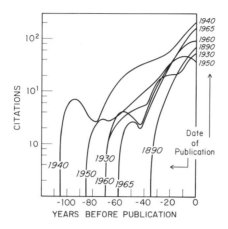

Fig. 6.2 Distribution of citations in earth science literature normalized to the same publication date.

old. As usual, geology in the nineteenth century proves to be a lively science.

In 1930, the half-life was a normal 15 years, and the total life was 70. By 1940, after decades of dormant geology, the half-life was longer than normal, and there was a secondary peak of very old citations. The references spanned 100 years, and many papers written between 1835 and 1855 were cited. The geologists who published in the same journal in 1890 saw no reason to cite anything older than 1855. By 1950, the literature was roughly normal with regard to citation-age distribution. Slow publication rates meant few citations less than 2 years old; the older literature had a normal half-life but extended back 85 years. In 1960, the half-life was less than 15 years, and no paper was more than 70 years old. The half-life was only 10 years in 1965, and the oldest paper was only 60 years, and thus at last there were no citations earlier than the twentieth century.

The whole history is comparable to the stretching of a rubber band (Figs. 6.2, 3.3). In 1890, it was slack. Interest in the literature of literature expanded as geology became dormant. The rubber band stretched back across an ever greater span of time and was taut in 1940. Then it gradually relaxed to the present.

JARGON

Geology students are confronted with long lists of special names of minerals, fossils, rocks, sedimentary layers, time intervals, and landforms. My wife once remarked that it sounded more like the classics than a science as I was memorizing a list of fossil names in Greek and Latin by reading them aloud. It seems hardly possible that this is a characteristic of a vigorous science. A subject which demands rote learning is rarely attractive to creative minds. Thus, a filter exists which tends to reject future research scholars and passes those who look upon memorizing as scholarship. A feedback results because if some special names are good, surely more names are better; consequently, the filter gets thicker with the passing of time.

It is always possible to take the infinite diversity of nature and the endless span of time and construct a system of more or less arbitrary divisions. A system, if simple, has the virtue that the divisions are logical and easy to remember. A system, if ideal, may even illuminate the most basic relationships or evolution of the components of a system. The periodic table of the elements forms such a system. Crystal structure and chemical composition likewise give a useful basis for a system of mineralogy.

Unfortunately, some things are too complex to permit simple systemization, and yet it is useful to have a system to facilitate discussion and ease comprehension. The vast variety of igneous rocks are a case in point. Color, composition, grain size, and mineral phases are among the more conspicuous variables. In the face of this diversity the custom of naming a rock after a type locality developed. A new rock type discovered at Tranquility Base on the moon presumably would be "tranquilitite." It is difficult to see how this procedure has been much help — a sentiment shared by some petrologists: "Every petrologist who has racked his brain to remember the meaning of bekinkinite or tsingtauite, leeuwfonteinite or uncompahgrite, will appreciate the humor of Gevers and Dunne who said in a recent paper that they had refrained from coining new names such as isandhlundhuluite and umkandandhluite for certain granodioritic rocks in Natal."[5] One system of rocks may be useful, but a system is organized for a purpose, and organizers may have dif-

---

[5] S. J. Shand, 1944, "The species concept in petrology," *American Journal of Science* 242, 48.

ferent purposes. Thus, a second system and a third may be proposed in good faith and with the best motives. Chaos inevitably ensues. A significant component of the scientific literature degenerates into the reorganization and adjustment and amendment of arbitrary systems which were initially of roughly equal merit.

These problems of systemization and classification are not confined to the igneous rocks. The complexity of sedimentary rocks is such that the same custom of naming them after type localities developed. Sedimentary rocks can be classified according to grain size, composition, age, fossil assemblages, variability, and so on. Most of these characteristics are gradational both vertically and horizontally. The potentiality for confusion is very great, and yet units of rocks must be identified in order to prepare geological maps.

Slowly and painfully geologists have identified and resolved many problems in classification. They may have paid a terrible price for their efforts. The time spent in organizing a system is not spent making observations or trying to understand the phenomena observed. The time spent in reorganization may be enough to induce a state of dormancy. Did this happen in geology? Did the effort of preparing, publishing, and reading classification and jargon amount to an important fraction of the whole? I have attempted a simple measurement of this effect by determining the fraction of jargon words at different times. I take "jargon" to include all taxonomic, stratigraphic, lithological, and mineralogical names as well as the special terms of geology. Counting is not very easy, because a geologist tends to forget which words are jargon, and a nongeologist does not know which common English words have special jargon meanings. Acknowledging, therefore, some uncertainty, I find about 2 percent of the *Bulletin of the Geological Society of America* was jargon in 1891 and 16 percent was jargon in 1941, according to a sampling of only a few thousand words. On the other hand, the qualitative differences are obvious. The bulletin of 1891 reads like English and a few minutes with a dictionary would make the whole comprehensible to anyone. The journal of 1941 is hard to believe. Much of our language consists of articles, pronouns, conjunctions, verbs, and the like. This means that the 16 percent of jargon constitutes most of the remainder, and that the whole is beyond comprehension by the nonspecialist and possibly by the specialist as well.

CONTROVERSY

A normal research paper in geology has a title such as "X a new Y from Z" for a new mineral or fossil description. Another characteristic type is "The origin of X," although these are rarer. We have found a sizable component of bibliographic literature in geological journals, and there are papers dealing with semantics as well. The titles include "Is the Boulder 'batholith' a laccolith?" and "The use of the term Pocono" and "Chemung is Portage" and "Meaningless versus significant terms in geological classification."

We turn our attention now to yet another class of papers with titles containing such words as "controversy," "problem," and "question." These words indicate a problem has been identified but usually that it has not been solved. Papers of this sort are typical of philosophy and the humanities. We may consider such subjects as the authorship of Shakespeare's plays or the origins of Christianity. Controversy of long standing arises whenever questions deemed significant cannot be solved. Amongst lawyers, philosophers, and some scholars a formalism for controversy has developed wherein it is acceptable practice to discuss arguments for and against something without any expectation of resolving the matter. This is less common in the physical sciences, although a formalism exists in operations research for exposing all possible sides of a question to view. Scientists generally turn their attention to problems that can be solved and thus have little occasion to write about those that cannot. Philosophers, on the other hand, may look upon solvable problems as trivial and unworthy of publication. If this distinction is just, the existence of formal controversy without expectation of resolution should not be characteristic of a normal growing science.[6]

Controversy has existed in geology from the beginning of modern times, simply because of the difficulty of agreeing upon arbitrary stratigraphic boundaries. Murchison mapping down through Silurian rocks and Sedgwick mapping up through Cambrian rocks in a different area could not agree upon the stratigraphic boundary between them. They tried to resolve these matters by visiting outcrops and discussing views in meetings of the Geological Society of London

---

[6] Price, "Citation measures of hard science." This contains a summary and analysis of different characteristics of science and humanities research.

Scientific Literature 143

from the early 1840's on. This gradually became a tedious and sterile debate which was only solved in 1879 by Lapworth, who inserted a demilitarized zone, the Ordovician System, between the Cambrian and Silurian. His paper is described by Woodward writing in 1907 in the following terms: "With the view of cutting the Gordian knot and of putting an end to 'the interminable discussion' . . ." [7] Thus, neither Woodward nor Lapworth, nor presumably their colleagues, either enjoyed this controversy or considered it to be a suitable subject for scientists. We may contrast this with a listing of "more important and distinctive events" of the period 1848–1888 as evaluated in 1932. Two items are "Discussion of the nature of 'Eozoon,' 1865–1894," and "Discussion of the 'Laramie problem,' age of the lignites, 1872–1897." [8]

These long controversies established a pattern in geology of essentially formal debate as a substitute for problem solving. Most of the debates consisted of discussions of origins and causes in the conclusions of research or review papers. The host of papers and books on the origin of atolls provides numerous examples of this genre. The great Darwin proposed the essentially correct hypothesis that atolls originate by subsidence of volcanic platforms. The great Reginald Daly introduced consideration of the effects of changing sea level and cooling during the ice age. Daly used the word "problem" in his title, but with the intention of solving it as soon as it was posed.[9] The great William Morris Davis titled a whole book *The coral reef problem* in 1928, but once again his objective was to solve a problem not just discuss it. Speaking of Darwin he says:

> It has been a great pleasure to try to secure for his early-framed theory of coral reefs, which after worldwide adoption between 1840 and 1870 was so strongly objected to and even rejected by a number of later observers between 1870 and 1910, the broader consideration that it so fully deserves.[10]

---

[7] H. B. Woodward, 1907, *The history of the Geological Society of London* (London, Geological Society of London), p. 187.

[8] H. L. Fairchild, 1932, *The Geological Society of America 1888–1930* (New York, Geological Society of America), p. 44.

[9] C. Darwin, 1842, "The structure and distribution of coral reefs" (reprint 1921, Berkeley, University of California Press); R. A. Daly, 1910, "Pleistocene glaciation and the coral reef problem," *American Journal of Science* 30, 297–308.

[10] W. M. Davis, 1928, *The coral reef problem* (New York, American Geographical Society), p. 548.

Davis has shrewdly pinpointed one aspect of the growth of formal controversy in geology. The pioneers of the nineteenth century identified big problems and, if they were solvable with the available tools, they solved them. The geologists who followed, presumably equally energetic and intelligent, were inevitably doomed to working on trivia until new tools were forged. One way to use up energy was to quibble on details of the pioneer's broad, synthesizing, and simple hypotheses. New hypotheses were built around new details of observations, and the forest became obscured by dense underbrush.

The great T. C. Chamberlin dealt with the dangers of leaping to conclusions in a classic paper of 1897.

> For a time these hastily born theories are likely to be held in a tentative way with some measure of candor or at least some self-illusion of candor. With this tentative spirit and measurable candor, the mind satisfies its moral sense and deceives itself with the thought that it is proceeding cautiously and impartially toward the goal of ultimate truth. It fails to recognize that no amount of provisional holding of a theory, no amount of application of the theory, so long as the study lacks in incisiveness and exhaustiveness, justifies an ultimate conviction.[11]

The basic problem of geology was that an increasingly large fraction of research did not and could not lead to incisive and exhaustive results.

Geologists developed notably flexible minds from trying to view all sides of all questions. This training successfully prepared them for working in chaos, but it resulted in more and more formal controversy. After a while the application of the method of multiple working hypotheses was sometimes replaced by the mere statement of various hypotheses without any resolution of the problem. An initial paragraph outlined each of the numerous published hypotheses relating some cause with an effect. Hypothesis number one was then discussed with a list of evidence which supported it and opposed it. The same procedure was followed for as many as half a dozen hypotheses. No hypothesis was rejected on the basis of new critical experiments or observations. None was shown to be quantitatively important or trivial by mathematical analysis; no tests were proposed. The end result was a burgeoning literature that was incapable of

---

[11] "The method of multiple working hypotheses," *Journal of Geology* 5 (1897), 837–848, quotation from p. 839.

Scientific Literature   145

forming the foundation for a further rapid advance. The next paper started a new foundation and so did the one after. The edifice of geology could not rise above the ground floor.

The discussion of controversies and problems and questions was not always solitary. Indeed, some academic wit appears in the publications of symposia. Take for example the following from 1948:

> In this symposium we are treating a problem that is really an old friend. This problem goes back through almost all the years during which there has been a science of geology. We have never satisfied ourselves as to the solution, and we furthermore must note that no very serious consequences arise if we do not arrive at that solution.[12]

Gradually understanding emerges in this way, but the point I am trying to make is that the process is far from normal science. Controversy and debate replace problem solving, and in doing so they halt the construction of the scientific edifice. It is evident that the debators would stop debating and start building if they had the means. The loss to the science is that they do not turn to another part of the edifice and build where it is possible. Near the end of the dormant period in geology, from 1940 to 1960, some seventeen different field and laboratory classifications of sandstones were proposed in the North American geological literature.[13] Thus, approximately once a year, somebody developed a classification, somebody wrote, somebody edited, somebodies reviewed, and many people read and tried to apply these classifications. Think of the effect.

LITERATURE AND DORMANT SCIENCE

We have examined some of the things that happen when the literature of science becomes dormant. The English deteriorates as concern with style grows. Jargon flourishes. An increasing fraction is literature of literature, and bibliographies grow longer, and citations grow older. These things have a pernicious effect in themselves because they simultaneously reflect and set the style of the science.

Consider only the effects of delays in publication time. The

---

[12] F. F. Osborne, 1948, in *Origin of granite*, Geological Society of America Memoir 28, p. 100.
[13] G. deV. Klein, 1963, "Analysis and review of sandstone classifications in the North American geological literature, 1940–1960," *Geological Society of America Bulletin* 74, 555–576.

Bulletin of the Geological Society of America usually published within two or three months after receipt of a manuscript in 1890 and 1910. Thus, one journal simultaneously offered the maximum in prestige, circulation, and speed. The full range of the earth sciences appeared in one place, and the speed minimized the possibility of duplicate effort and resultant battles about priority. By the 1950's the publication time was about a year and a half, and two years was not unknown. Such delays are intolerable in an active field in which the whole literature may double in four years. At first the delays were accepted to achieve wide circulation and maximum prestige. However, this resulted in duplication of effort and overlapping publication, so that the same results were in press by more than one person. A confused terminology and priority battles followed. Earth scientists working in active fields published increasingly in *Science* and *Nature* and the *Journal of Geophysical Research*. Classical geology became more isolated and withdrawn just when a reawakening became possible.

The situation was no better, perhaps worse, in society meetings. The International Geological Congress demanded abstracts and often manuscripts perhaps two years ahead of a meeting. This made sense in fields where long experience had shown that only one man in the world worked on a given subject. It seemed reasonable to people who submitted reviews or new classification schemes and who expected formal controversy and debate. Anything written that far ahead in an active field, however, would be out of date by the time the talk was given. Consequently, many of the newer earth scientists either talked about trivia or did not go. This applied as well to the meetings of the Geological Society. Abstracts were due roughly a year before the meeting, and they had to be submitted on exactly the right form to be accepted. Such bureaucracy not only excluded isotope geochemists and marine geophysicists from the meetings, but even kept them out of the society. A further splintering of once monolithic geology ensued. Geologists refused to recognize that a man working on a ship might be one of them. Geochemists in response refused to be called "geologists" and took to the term "earth scientist." I do not think it is excessive to blame this split on the long dormancy of geology.

If one science, why not all? If American science in general is stifled and becomes dormant, it seems reasonable to anticipate that

Scientific Literature 147

the scientific literature will take on the character of the dormant period in geology. The cost may be substantial.

SCIENTIFIC PUBLICATIONS OF THE FEDERAL GOVERNMENT

We can consider one aspect of the cost by considering what happens to budgets in government during periods of variable growth. The scientists in government service work under different conditions than those in universities in that they are parts of a vast bureaucracy. Nevertheless, we have seen that they are affected in the same way as other scientists by variations in growth rates. Morover, government records are more easily acquired than those of universities, and they lend themselves to types of correlations otherwise impossible. For this purpose we can hardly do better than to study the United States Geological Survey, because it is generally regarded as among the premier scientific agencies in the world and it publishes voluminously — which amounts to the same thing. Whatever conclusions emerge, we shall be justified in believing that they represent the optimum for federal science and that other agencies would not have such favorable records.

Let me stress the point. When planning a federal reorganization, for example, the role of the Geological Survey is often designated as "increasing the level of scientific competence in the new agency." As to individuals, the survey has had eight directors, and six of them have been members of the National Academy of Sciences. Equally distinguished scientists have been associated with the organization throughout its long history.[14] One reason is because of such administrative innovations as a policy of rotating creative scientists into upper administrative posts and then back into research if they wish. As a consequence, a scientist can be temporarily attracted to administration and promotion, whereas in other agencies he would either be underpaid or stop doing research. The survey has also maintained a free interchange of its personnel and problems with the outside scientific world. Scientists not infrequently work in industry or universities as part of a predominantly survey career. Professors and students may work part time on survey problems "When Actually Employed." This WAE device provides consultants at the wages of regular workers and thus is one of the most econom-

[14] In 1885, even the librarian was named Charles Darwin.

ical ways of doing science yet discovered. It was very popular with academicians as a salary supplement that could not be obtained in any equally desirable way earlier in the century. Now it tends to be replaced by the far more expensive research grant from the Navy or National Science Foundation.

The history of the Survey is generously documented and universally praised.[15] John Allen of Portland State College has observed that what we know as "Parkinson's Law" was stated as early as 1885 by Clarence Dutton in referring to the survey of which he was a member.

> Our Survey is now at its zenith and I prophesy its decline. The "organization" is rapidly "perfecting," i.e., more clerks, more rules, more red tape, less freedom of movement, less discretion on the part of the geologists and less out-turn of scientific products. This is inevitable. It is the law of nature and can no more be stopped than the growth and decadence of the human body.[16]

In order to test Dutton's prediction, we have available data on input in the form of funding and staffing of the survey and on output in the form of pages of scientific papers. I pause only to emphasize that this is possible because the survey *has* an output of this type. Most agencies including those concerned with the environment and science, do not.

Figure 6.3 shows the number of scientists, employees of all types, cumulative expenditures, and cumulative pages published by the Geological Survey. There are about four times as many nonscientists as scientists, and both reached peaks in 1920 and then dropped 40–50 percent before resuming the long-term rise which brought full-time employees to almost 8,000 by 1965. The number of scientists was only 200 in 1920 and did not reach that number again until 1940. During the same interval other employees almost doubled.

The pages of scientific output, as we know, increased very rapidly until 1910 and then began to slow. From 1925 to 1955 the growth was exponential with a doubling time of 70 years. From 1920 to 1955 was the dormant period in American geology, and during that time survey publications increased by only 20 percent. During the

---

[15] T. G. Manning, 1967, *Government in science* (Lexington, University of Kentucky Press).

[16] G. Y. Craig and M. C. Rabbitt, 1969, "Letters from American geologists," *Geotimes 14*, no. 5, 21.

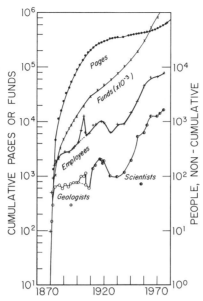

Fig. 6.3 The growth of various components of the Geological Survey. The plots of pages and funds are cumulative, but those of total employees and scientists are not.

same 35 years, scientists and total employees each increased by 570 percent, and costs went up 1,000 percent. The cumulative budget divided by cumulative pages of publication can be used as a measure of cost per page. This varied only from $160 to $200 from 1880 to 1920. By 1935, however, it was $330 per page, and by 1955 it was $1,000 per page. The inflation continues because costs are rising faster than pages of output, even though the latter have been accelerating since 1955. In 1965 it cost $1,600 per page.

Inflation has its effects, and the Geological Survey does other things than publish pages. Nevertheless, Dutton's prediction is confirmed in that the size of the budget and the staff are unrelated to the output of science. This corresponds to Parkinson's demonstration that the admiralty staff increases without regard to the size of the operational fleet.[17] If Parkinson's laws apply to the best on land and sea, is there any way for governments to escape them? The individual states of the Union did so for 80 years by refusing to play the game. With regard to growth rates of science, the issue is clouded because

[17] C. N. Parkinson, 1957, *Parkinson's law* (Boston, Houghton Mifflin Co.), p. 7.

the records do not overlap the dormant period in geology. Nevertheless, it was at the very least an administrative triumph and worthy of examination.

LITERATURE OF STATE GEOLOGICAL SURVEYS

The history of the state geological surveys provides remarkably useful examples of the interaction of government with the environmental sciences. In the first place, so many have existed and over such a period of time as to compose a splendid sample. Moreover, the history in terms of people, costs, legislation, and output is readily available from questionnaires submitted by the surveys and compiled and analyzed by Merrill.[18] The general style and objectives of the state surveys are indicated by the wording of the first enabling act passed by the legislature of North Carolina in 1828:

> ... employ some persons of competent skill and science, to commence and carry on a geological and mineralogical survey of the various regions of this State; and that the person or persons so employed shall, at stated periods, furnish to the board true and correct accounts of the results of said surveys and investigations, which shall annually be published by the board ...
> ... a sum of not exceeding $250 be, and the same is hereby, annually appropriated for four successive years . . .[19]

Thus the survey was to be conducted by skilled men, it had a discrete limit of time and costs, and the results were to be published.

The state surveys began before the formation of the Geological Survey of Great Britain in 1835, and they flourished before the formal organization of the U.S. Geological Survey in 1879. One survey was active in 1830 and twelve in 1840, but during the interval five had been created, had done the work agreed upon, and terminated. Because of the constant coming and going, a total of 40 states had had geological surveys by 1900, but there were only 3 in 1850, 14 in 1860, and 13 in 1870. In these early times the required professional geolo-

---

[18] G. P. Merrill, 1920, "Contributions to a history of American State geological and natural history surveys," *U.S. National Museum Bulletin 109*; 1924, *The first one hundred years of American geology* (New Haven, Yale University Press; facsimile reprint, 1964, New York, Hafner Publishing Co.).

[19] Merrill, "Contributions to a history of American State geological and natural history surveys," p. 366.

gists would not have been available to staff all the surveys at once. Instead, a team conducted a survey and then commonly transferred as a team to another state and did another. It is not apparent that these were formally organized teams (I think they were not), but the same names appear in the same organizational arrangements in one survey after another. Indeed, the surveys collectively lasted long enough for families of professional geologists to be active. W. B. Rogers, for example, was assisted in the geological survey of Virginia in 1835 by J. B. Rogers, H. D. Rogers, R. E. Rogers, and others.[20] It appears that he and Virginia were fortunate, because H. D. Rogers went on to be State Geologist of New Jersey and of Pennsylvania, and three of the Rogers brothers were founding members of the National Academy of Sciences.

Men of this quality were common in these surveys, and they were vigorous in meeting the requirements to publish the results. We have already seen in Chapter 3 that the outpouring of geology in state survey papers surpassed scientific journal publications by 1860 and has never been equaled by the U.S. Geological Survey. The cost of the publications reached $100,000 by 1850 and $1,000,000 by 1860, which are not enormous sums even allowing for inflation. However, to obtain a realistic picture of the magnitude of the effort, we can compare the annual expenditures with those of the federal government. In 1860 the budget of the United States was about $60 million and that of the state geological surveys was roughly $60 thousand, or 0.1 percent. In 1965 the federal budget was about $100 billion, and the same percentage would give $100 million for state geology. This does not seem very great considering that more than $1,500 million was spent by the states for natural resources in 1963. In sum, the surveys of the nineteenth century were low-budget operations.[21]

The cumulative costs of all the surveys can be compared with the output of pages of reports as we did for the U.S. Geological Survey. By this measure the states did much better. The cost of a page was less than $10 up to 1840, but by 1850 it had risen to $30. The reason for this rise was one of the first outbreaks of deliberate defiance of government policy by a scientific agency, namely, the geological sur-

[20] Merrill, *The first one hundred years of American geology*, p. 183.
[21] U.S. Bureau of the Census, 1965, *Statistical abstract of the United States: 1965* (Washington, Government Printing Office).

vey of the State of New York. This provides a nice example of the wealth of information available on the interaction of science and government.

The legislature appropriated $104,000 in 1836 to be spent for four years of geological surveying leading to a comprehensive report. By 1840, only $72,882.12 had been spent, and a continuation of the survey without additional funding beyond the original amount was authorized. The accounting to the penny seems ironical in that

> . . . through an error in bookkeeping it later developed that this amount had already been exceeded by $228.80, with considerable still due on salaries and publication of reports as yet unprovided for. It was then estimated that the sum of $45,363.90 additional would be required.[22]

By 1844 the legislature was under the impression that somehow the total cost had gone up to $235,902, but the select committee appointed in 1849 to investigate the survey discovered the total exceeded $425,000. James Hall was called upon to explain why so much money had been spent on publications, including the following:

Original drawings	$ 8,360
steel engravings	9,871
stone engravings	16,818
wood engravings	20,363
printing, letterpress, and binding	89,072
coloring	73,212
printing impressions of plates	38,582

The legislators thought that this was unreasonable. The scientists thought the legislators were unreasonable. Louis Agassiz wrote to Hall in 1849 when the investigation began:

> In the first place, let me remark in general manner that it is deeply to be regretted that with the most liberal dispositions legislative bodies and governments scarcely ever understand the wants of science; and having no opportunities of intercourse with men of science, I do not mean professional men in scientific professions, but men of original research, they can not understand how science can be promoted, and make often the greatest blunders with the best intentions. It is unpleasant to say, but it is so, and unless you can make your people understand that no investigations can be hurried, you will never have independent in-

---

[22] Merrill, "Contributions to a history of American State geological and natural history surveys," pp. 335–336.

Scientific Literature 153

vestigators in this country, and the few who prefer their scientific reputation to any position in society will be left to struggle with never-ending difficulties.[23]

The legislators had appropriated the money for the promotion of the welfare and development of the resources of the State of New York, however, not for the promotion of science or scientists. The select committee sounds vexed in its report:

> 2. The whole original plan of publication was departed from, instead of 3 volumes octavo there are to be 20 volumes quarto. In 1842, when the quarto size was determined on, it was supposed eight volumes would be the number.
> 3. The addition from time to time of new departments to the work. Paleontology and agriculture alone add eight volumes not originally contemplated.
> 4. In the increase of illustrations and the amount of coloring will be found the great cause of expense.[24]

Hall himself wrote to the committee in response to these charges before the report was issued. The main emphasis of his letter is on the number of plates being an absolute minimum for the purpose. How then do we reconcile this with the findings of the committee as reported by Merrill?

> The work grew in proportion to the opportunities offered, however, and finally the survey was the subject of a legislative investigation with the usual disastrous results. There was no question but that good work was being done, but, as it seemed to the layman, overdone, particularly when the tendency to super-illustrate certain species of fossils was considered. It was shown in one instance that 8 species of fossils had been illustrated by 117 figures, one *Spirifera radiolarius*, having been figured 27 times, the drawings and engravings for which cost $526.50. Again, it was shown that the fossil *Pentamerus galeatus* was pictured by each of the four geologists and by a total of 74 figures, the drawings and engravings for which cost $174, in addition to the $150 for printing, or $324 for the one species.
> It would be interesting to note how many times and in how many works these same species have since been figured.[25]

Merrill's final comment suggests that he believed that Hall was in the right in this matter. The legislature thought not and appealed for tighter controls and continuity of management.

[23] Ibid., p. 337.
[24] Ibid., p. 334.
[25] Merrill, *The first one hundred years of American geology*, p. 232.

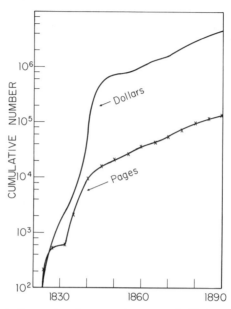

Fig. 6.4 The cumulative cost and output in pages of all state geological surveys.

Control and continuity are what we have in the federal government, if anywhere. In that paragon of government agencies, the Geological Survey, the average budget to yield a page of report was about $160 at the beginning, and much more than $1,000 80 years later. What happened in the loosely administered, organizational shambles of the state surveys? The cost before Hall's battles with the legislature was less than $10 per page, and then it jumped to $30. What is probably far more important is that it stayed at that cost for the next 40 years (Fig. 6.4). Through some miracle the growth of pages was proportional to that of costs despite inflation, the civil war, the westward expansion, and despite everything else. Parkinson's Law was repealed.

GOVERNMENT SCIENCE AND GROWTH RATES

The difference between the state geological surveys and the U.S. Geological Survey is not in the quality of personnel, because the same people were active in both. It is the result, rather, of some effect of scale or permanence. More probably it is the latter, because permanent agencies of all sorts appear to follow Parkinson's Laws regardless of size. We deduce therefore that temporary scientific agencies are efficient and permanent ones are not. Once again, a scenario is not difficult to compose. The temporary agency has a job to do and a budget, with only a limited prospect of much expansion of either one. Little time need be expended in committees evaluating prospective new staff members. Meetings of the committee to plan a large permanent laboratory building for the agency can be dispensed with altogether. This forestalls another of Parkinson's Laws: that the utility of an organization vanishes as soon as it moves into specially designed quarters.[26]

We may contrast a temporary agency with the typical permanent agency, which in fact is in the special laboratory buildings that were built twenty years before to make it more efficient. It is not solving the assigned problems, either because it is incapable of doing so, or because it is highly capable and has long since done so. In any event, it is a drain on department or agency efficiency. Costs rise steadily as new staff accumulate, but output vanishes.

What is to be done? It is not humane nor even politically feasible to fire everyone who is inefficient or close every permanent agency and thereby make it temporary. The same scientists are inefficient in one administrative arrangement and efficient in another. How do we put them in the latter? I believe it is by emphasizing problem-solving and by periodic retraining so people remain able to solve new problems. We are confronted with vast organizations assembled to develop rigging for clipper ships or triggers for atomic bombs, organizations that did these things and continue to do them because they don't know how to do anything else. The organization needs a new mission to be effective, but no one knows what to do with the workers who are skilled only in the old mission.

We have seen what happens. We need to create temporary agencies to solve problems, and we end up with permanent ones for supposed

[26] Queen Elizabeth, with her unerring instinct for economy, also forestalled Parkinson's Law even in the operation of the British Navy. In the days of the Spanish Armada there was no permanent Admiralty or fighting fleet. The whole thing was temporary. As many as possible of the captains and ships were kept gainfully employed in the very model of private enterprise, namely, privateering. When needed, Lord Howard was designated Admiral, and Drake was called to active duty from his temporary pursuit of bowling. After the Armada was defeated, they were naturally taken off the payroll. The administrative policy had become much more generous by Nelson's time, and unneeded captains were given half-pay for standing by. Although these policies would hardly survive labor negotiations now, it is well to remember that they were in force when sea power was at its peak.

humanitarian reasons. Actually it is not very kind to doom competent people to futility when they could be doing something useful. Let us say we declare the original mission of the agency is solved, send everyone back to school, and start on a new mission. Life is full of excitement and meaning. Presumably this rather obvious solution has not been tried because of the supposed expense. However, compare the costs for a page of report from permanent and from impermanent agencies. We may generously suppose that the costs to produce a page will be much greater for a large agency than for a small one. Thus we merely compare the increase in costs. For a large permanent agency the average cost went up eightfold after an initial period of stability. For a group of temporary agencies there was no increase at all. We may surmise that it costs no more to reeducate people than to employ them full time. In this example, most of the people in the permanent agency could have been retraining most of the time if only when working they were as efficient as in the temporary agencies.

7 Education

SUPPLY AND DEMAND OF SCIENTISTS

The only significant source of scientists and technologists is the universities of the world. The universities supply and the employers demand. In a perhaps limited sense, education is an industry — subject to analysis like other industries — and prey to the business cycle. For our purposes this may be generalized as follows:

1. For the long run, supply equals demand.
2. If demand temporarily exceeds supply and is growing, the supplier expands plant facilities and staff to meet anticipated as well as existing demand.
3. A boom exists and continues until capacity equals or exceeds demand.
4. Alternatively, the supply of money for expansion is exhausted or becomes prohibitively expensive in a credit pinch.
5. During a boom, inventories are low.
6. If required capacity is overestimated, a surplus develops which initially goes into expanding inventory.
7. Unless balance is achieved smoothly, a recession ensues.
8. This is marked by idle plant, unemployment, and a continuing surplus of supply over demand.

The intensity of the cycle depends on the nature of the industry and the extent and success of government countermeasures designed to subdue it. It is presently negligible for an industry such as an electrical utility, but that can change, as it did for railroads. An intense cycle is characteristic for heavy industries such as steel production and automobile manufacturing. The remaining type of industry is identified by rapid growth in which the rate varies, but not the constant upward trend. But, as many a burned speculator knows, in the initial phase of expansion a cyclical industry is indistinguishable from a growth industry. This makes the initial recession all the more unexpected and the cyclical effects all the more pronounced. If education is an industry, it is clearly very important to identify its type and the stage of the cycle that it is in.

Professors do not consider themselves in an industry, and students do not regard themselves as potential "output" of a university nor as potential "input" to employers. Education is a different kind of activity I am told; and as a professor, I like to believe it. Universities are engaged in the pursuit of excellence; can they be a business? Some plants in the aerospace industry are committed to absolutely error-free manufacture; are they universities?

Let us make a preliminary sketch of the history of the education of scientists for the last few decades and see what has happened in this activity that is different from industry. After the Second World War scientists were in demand, and more students majored in science. The federal agencies poured money into universities, which built new laboratories and classrooms and hired new professors — often at premium salaries. Exponential growth ensued and requirements for further staff and facilities were predicted by assuming continuing expansion. For various social and economic reasons the expansion stopped. In California, at least, it is partially because the authorized state bonds cannot be sold at legally permissible interest rates. Inasmuch as universities absorb many new doctorate recipients, some had difficulty in finding jobs. In 1959, when jobs were plentiful for new physicists, only 6 percent took positions in the pleasant limbo of postdoctoral appointments. By 1967, the fraction was 25 percent, and in 1970, it was 46 percent — mostly in temporary positions created by chairmen of the departments that trained the graduates.[1]

It was difficult to write the last paragraph without using the words "plant expansion," "boom," "credit pinch," and "inventory buildup." If education is not an ordinary industry, it nevertheless seems to behave in ordinary ways. Thus, it does seem worthwhile to study this activity which seems to train scientists with little regard for demand for their services, and moreover without the students' having much regard either. We seek to find its general character, its support, the variables that influence it, whether it has adequate or

[1] P. H. Abelson, 1970, "The changing job market," editorial in *Science 168*, no. 3933. Data from Susanne D. Ellis.

Education

excess capacity, and other factors that will guide predictions of future growth.

THE CHARACTER OF THE INDUSTRY OF EDUCATING SCIENTISTS

The annual production of baccalaureates in the United States has doubled on average every 15 years since 1900, even including the world wars. Annual awards of doctorates in all fields have doubled every 10 years during the same period, and the fraction in the sciences has remained relatively constant at least from 1920 to 1960.[2]

These facts are enough to demonstrate that the output of education has grown with the same persistence and at about the same rate as, for example, electrical generating capacity. Output of each is expanding faster than the general growth of the population. More people are getting more education, as well as more electricity.

Output can be increased by capital investment to make labor more productive or by increasing the labor force and maintaining individual productivity. Electrical utilities invest capital; so do universities, but to a large extent they have increased output by increasing staff — which requires less capital but more continuing expense for salaries.

Another characteristic of education is that it loses money on each output, that is, the students do not pay fees equal to the cost of their education. This might be taken as grounds for considering university education of scientists as a public service. However, science students are trained for jobs as scientists, and if there are no jobs, the end result is hardly a public service. It seems at least as valid to regard education as a subsidized business like shipping or railroads or farming and so on. If so, the question that arises is, what is the nature of the subsidy?

To attempt to understand the education of scientists, we can compare degree output with social and political factors and other variables such as federal funding. In doing this we should not lose sight of the most important characteristic of this education, namely, that it

[2] U.S. Bureau of the Census, 1958 and annually, *Historical statistics of the United States, colonial times to 1957, and continuations* (Washington, Government Printing Office); National Academy of Sciences, 1963, *Doctorate production in United States universities, 1920–1962* (Washington, National Academy of Sciences); National Academy of Sciences, 1967, *Doctorate recipients from United States universities, 1958–1966* (Washington, National Academy of Sciences).

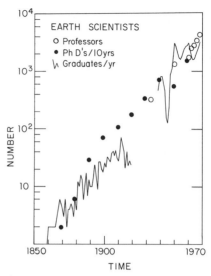

Fig. 7.1 The annual output of graduates, dicennial output of doctorates, and standing population of professors in the earth sciences. The plots are not cumulative.

has long remained proportional to all other higher education. Thus, we must seek agents that expand degrees in nonsciences as much as in sciences. It may be quite true that a drop in special fellowship support for science students would reduce the number of such students relative to other fields. Nevertheless, the reverse has not been true. A massive input of special fellowships for scientists has not caused any significant increase in the proportion of science students.

TRENDS, FLUCTUATIONS, AND THEIR CAUSES

A long time series related to education output can be constructed from the baccalaureates in geology, although there is some uncertainty in the results. Data exist from 1857 to 1919 in a single series (Fig. 7.1).[3] The next census of graduates in the earth sciences was undertaken by A. I. Levorsen, who polled first 64 and then 75 colleges beginning in 1938.[4] The sample consisted of the number of

[3] E. B. Mathews and H. P. Little, 1921, "Geology and geography in the United States," *National Research Council, Reprint and Circular Series*, No. 17.

[4] A. I. Levorsen, 1949, "Survey of college students majoring in geology," *Bulletin of the American Association of Petroleum Geologists 33*, 1928.

Education

seniors at major colleges only. Most seniors graduate, so little error results from taking the two as equal. The effect of the sample size is suggested by a comparison with Office of Education data that became available in 1949; the poll gave 1,715 seniors and the Office of Education gave 1,850 recipients of a B.S. or B.A. in geology. A possible error of 10 percent is indicated in this historical census. Levorsen's data show seniors increasing from 470 in 1938 to 812 in 1940 and then dropping to 129 in 1945. After the war the number of seniors increased very rapidly to 2,623 in 1950, according to this poll. Graduates totaled 3,043 according to the Office of Educaton.[5] Another major perturbation followed; output varied from 3,043 degrees in 1950 to 1,632 in 1953 to 2,816 in 1959.

For the decade of the 1960's, data are readily available on the number of seniors majoring in the earth sciences.[6] They fluctuated from 3,200 in 1960 to 1,600 in 1963 and back up to 3,300 in 1968.

In summary, there is every evidence that the average annual number of graduates in the earth sciences has both increased and fluctuated. This makes it difficult even to estimate what happened during the 1920's and the early 1930's, when the average effect was an increase from 5 degrees in 1920 to at least 470 in 1938, or almost two orders of magnitude. For want of a more rational plan, I have assumed that the number of graduates in 1921 returned to an average figure of 45, which was typical of the prewar years. After that I assume a simple exponential expansion to 1938. If so, the total cumulative output of graduates in geology passed 100 in 1883, 1,000 in 1920, 10,000 in 1946, and is now roughly 60,000. What was the cause of the fluctuations? Much of the literature about the graduates deals with relatively short intervals in which the fluctuations loom large and require some explanation. Thus, Mathews and Little considered the then large number of 72 graduates of 1912 as "entirely out of proportion and represents a marked departure from normality."[7] By 1938 about 460 geologists graduated per year. Likewise, the marked decrease in graduates after 1912 was attributed as follows: "The recent excessive demand for geologists in applied geology

[5] C. B. Lindquist, 1961, "Geology degrees during the decade of the fifties," *Geotimes* 5, no. 8, 15.
[6] G. R. Downs and B. C. Henderson, 1968, "Student enrollment in earth science, 1967–68," *Geotimes* 13, no. 9, 20–21.
[7] Mathews and Little, "Geology and geography in the United States," p. 4.

has drawn students away from the universities before their training has been completed." [8] Just the opposite explanation was advanced for the drop in graduates in the early 1950's and again in the early 1960's. At both times industry, mainly oil companies, sharply reduced recruiting or actually fired geologists. The decline in graduations then occurred after an interval adequate for undergraduates to have shifted to another major. The correlation seems convincing, although it appears that oil compay recruiting is the whipping boy in explaining fluctuations up or down. It might appear, alternatively, that the decline in the 1950's was an effect of the third major American war of the century, but the number of women graduates, unaffected by the draft, fluctuated almost the same way.

If contemporary observers correlate opposite causes with the same effect, it is always possible that the phenomenon is essentially random. This is well known in the stock market, in which random fluctuations of market indexes are always explained in terms of the news of the day without regard for all previous explanations. Thus, on Monday, "The market rose in response to rumors of peace." On Tuesday, "Overoptimism on Monday caused the market to fall, despite continuing rumors of peace." Let us consider the fluctuations of the annual number of graduates just as fluctuations and see if we come to any conclusions. A general rising trend has existed for 100 years. The limits of fluctuation from 1860 to 1920 were confined to a linear band, with the average values doubling in about 12 years. The width of the band was about 4 times the minimum value at a given time. Within the band are a series of steps with a period of about a decade. Rapid rises were brief and were followed by a relatively constant or even decreasing number of graduations in the 1860's, 1870's, 1880's, and 1910's. Thus they were typical.

After 1938 no pattern of growth is obvious. Even during wartime, graduations never dropped below the extended minimum trend line of the earlier period. However, at most times the graduations were well above the top of the older trend channel. The decade-long fluctuations seem to be continuing. During wartime the maximum fluctuation was 6 times the minimum value or larger than earlier. Since 1950, on the other hand, the fluctuations have been only twofold or much less than at any time in the past. The small percentage

[8] Ibid., p. 21.

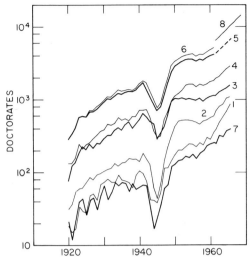

Fig. 7.2 Doctorates in sciences and engineering (NAS 1142, NSF 69-37, NAS 1489 [1967]). (1) Mathematics; (2) Physics; (3) Chemistry; (4) "Bio-sciences"; (5) Σ Natural Sciences; (6) Σ Natural Sciences plus engineering; (7) "Geo-sciences"; (8) Science and engineering NSF 69-37.

of variation contrasts with the numerical variation, which is the largest on record.

This analysis tells us that the two- to eightfold fluctuations of the past should be considered as normal rather than as a cause for surprise. They may be expected in the future unless something is done to alter the factors which produce them. It seems hard to believe that all these fluctuations were responses to the same cause as was the case recently, namely, variations in the employment practices of industry.

From this historical census, it appears that the output of the education system varies in time. Although its average growth resembles an electrical utility, it fluctuates more like a cyclical one.

DOCTORATE TRENDS

We can draw upon data on doctorates in all sciences for further information about trends and fluctuations in output and their cause. The annual output of doctorates in science increased from 279 in 1920 to about 6,800 in 1966 (Fig. 7.2). The total for science and engineering was 14,100 in 1968 or almost equal to the total for the interval 1920 to 1940.[9] The time series is much shorter than for geology baccalaureates, but it shows comparable fluctuations, perhaps because it spans a major depression, a great war, and considerable variations in the support of science. Annual doctorate output in each of the sciences approximately quadrupled in the 1920's, and not one even doubled in the 1930's. Indeed the total annual number of doctorates in the sciences increased by only 50 percent in that decade, and geology and mathematics grew hardly at all. The 1940's with World War II saw a spurt in doctorates in 1941–1942 as students finished while they could. This was followed by a marked drop by 1945, but each of the sciences finished the decade with output far ahead of the beginning. How far ahead seemed related although not proportional to the depth of the wartime drop. Biology, chemistry, and mathematics each dipped about 60 percent by 1945, but through the whole decade they expanded 70–100 percent. Geology and physics plummeted 73 percent and 70 percent, but expanded 123 percent and 193 percent respectively during the decade. The drop and rapid expansion of physics are readily accounted for in the need for physical scientists in war research, the academic development of classified research in atomic physics, and so on. The causes of the changes in geology are more obscure. Perhaps as the smallest population it was subject to relatively large fluctuations.

In the 1950's, the annual output of doctorates in science increased from about 3,000 to 4,000; this was the slowest pace for any decade for which comparable records are available and certainly since 1920. Again the individual sciences had variable growths. Chemistry and physics hardly varied at all. Biology increased for the first few years of the decade and then remained relatively constant. Mathematics almost doubled, and geology did so, to continue the relatively rapid growth it experienced in the 1940's.

Consistent data are available through 1968, and we can assume that the remarkably linear trends continued through the decade (Fig. 7.2). It was a period of extraordinary growth, unmatched since the 1920's, when the absolute numbers were an order of magnitude smaller. Doctorate output in most sciences doubled or tripled; the

[9] National Science Foundation, 1969, *Science and engineering doctorate supply and utilization, 1969*, NSF 69–37 (Washington, Government Printing Office).

Education

only exception was in mathematics, which expanded sevenfold. The sciences as a whole, however, did not increase relative to nonsciences. So much for the post-Sputnik spurt.

Doctorates in the sciences during the last half-century, like the output of scientific papers, have fluctuated in complex ways and presumably in response to complex influences. Very rapid expansion occurred in the 1920's for no obvious reason except that the nation was prosperous. Expansion slowed in the 1930's for the obvious reason that there was a depression. It fluctuated widely in the 1940's, but the net effect of an unprecedented war was not as great as the depression when spread over a decade. Indeed, the wartime infusion of scientific and technological sophistication into industry may have been one of the principal causes of the subsequent growth. We can hardly imagine the rapid expansion of doctorates in atomic and nuclear physics without the billions of dollars poured into the atomic bomb and the Atomic Energy Commission. The other great cause of expansion was the GI Bill of Rights, whereby the federal government paid for veterans' education. I join the other beneficiaries in restating the enormous impact of this program. Almost every graduate student in geology directly after World War II was a veteran, and this was probably true of all other sciences and of the nonsciences that increased proportionately.

The 1950's saw a slow expansion little different during a time of broad prosperity from the pace in the Great Depression. This is enough to raise questions about the "normal" rate of output of scientific doctorates. Immediately after Sputnik in 1958, the government began a program to subsidize the education of scientists. It was vast, but it affected by no means as large a fraction of students as the support of the GI Bill in the late 1940's. Except in these two periods of direct subsidy, the annual output of doctorates in science has increased ever more slowly. Perhaps we were not very far from the ultimate limit of the fraction of the population capable of and interested in obtaining scientific doctorates in 1958. The nation is nowhere near granting doctorates to all those intelligent enough to earn them in science, but it may be approaching the limit of those who want to do so without special inducements. Students have been willing to undergo the rigors of scientific training while generously supported by federal fellowships and with the glittering prospect before them of good jobs. There is little reason to hope that as many students will continue to do so while the support and the prospects melt away.

CAPACITY OF THE UNIVERSITY SYSTEM TO EDUCATE MORE
UNDERGRADUATE SCIENTISTS

A matter of great importance in evaluating the health of an industry and in forecasting events is an assessment of plant capacity. Is the industry capable of meeting demand or, indeed, does it have an excess capacity? For education we can gain intimations of an answer by comparing present capacity with that of the past and by constructing a model of expectable output and comparing it with real output.

Present and past capacity can be compared by evaluation of the ratio of faculty to undergraduate students at different times. As usual, data are available from geology. During the last two decades the annual number of graduates in geology has remained relatively constant (Fig. 7.3), merely oscillating around a horizontal axis. Meanwhile the number of professors of geology has grown with an average doubling time of 10 years since 1933, if we can compare measurements from different sources. A relatively consistent series of polls from 1960 through 1968 indicate a doubling time of only 7 years for that period.[10] However, Figure 7.3 leaves little ground for uncertainty about the general trend of the faculty-student ratio. From 1950 to 1968, the annual number of graduates increased only very slightly from 3,100 to 3,300. Meanwhile, the 1,100 professors of 1950 increased to more than 4,000 in 1968. We can now perceive a basic flaw in Martino's otherwise useful analysis relating the need for professors of science as a function of need for graduates in science.[11] He assumes that some relationship exists whereas our data indicate the contrary. Within broad limits the number of professors is unrelated to the number of graduates. I feel justified in generalizing from geology to science as a whole because in recent years the number

---

[10] American Geological Institute Committee on Geological Personnel, 1950, "Supply and demand for geologists, 1949–1950," *Bulletin of the American Association of Petroleum Geologists 34*, 1934–1942; H. R. Fairbanks, 1936, "Geologists: their distribution and background," *Proceedings Geological Society of America for 1935*, pp. 443–468; B. C. Henderson, 1969, "Manpower and salaries," *Geotimes 14*, no. 2, 13–19.

[11] J. P. Martino, 1969, "Science and society in equilibrium," *Science 165*, 769–772.

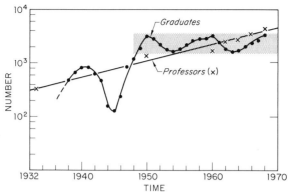

Fig. 7.3 The number of professors increases without regard for the number of graduates in the earth sciences.

of physics professors also increased rapidly while undergraduate majors were relatively constant.[12]

With regard to plant capacity, there is no record that geology professors were overloaded in 1950. Inasmuch as there are now four times as many professors and no more students, we have a suggestion, a hint at least, of surplus capacity. Either the input of professorial effort is quantitatively unrelated to the educational output or else professors of science spend most of their time in efforts other than educating undergraduates to be scientists. Without doubt they could quickly begin to train more should the need arise.

CAPACITY TO EDUCATE GRADUATE STUDENTS

The capacity of the universities for education in graduate science can be evaluated by comparing it with various growth and output models. The growth of doctoral degrees in subfields of geology provides examples of the factors that shape graduate education in the sciences. We shall first elucidate the nature of this growth and then turn to models to explain it and their implication.

The titles of all 11,091 geological theses accepted for advanced degrees in the United States and Canada during the century ending in 1957 are tabulated by Chronic and Chronic.[13] They have been

[12] Abelson, "The changing job market."
[13] J. Chronic and H. Chronic, 1958, *Bibliography of theses* (Boulder, Pruett Press, Inc.).

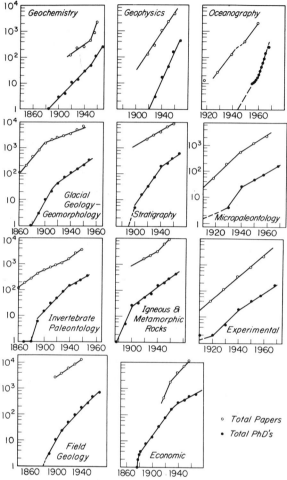

Fig. 7.4 Growth of doctorates and papers in subfields of the earth sciences.

classified into subfields and tabulated by date in order to analyze growth rates.[14] The growth of doctoral degrees in eleven subfields

[14] The categories are as follows: (1) economic, nonpetroliferous; (2) economic, petroleum; (3) engineering, water; (4) experimental; (5) field, quadrangle; (6) regional, historical, field technique; (7) geochemistry; (8) geomorphology, glacial, Pleistocene; (9) geophysics, seismic; (10) geophysics, nonseismic; (11) igneous, vulcanism; (12) marine; (13) metamorphism, Precambrian; (14) mineralogy, crystallography; (15) paleontology, invertebrate and botany; (16) paleontology, microscopic; (17) paleontology, vertebrate; (18) petrography, petrology; (19) sedimentation; (20) stratigraphy; (21) structural. These were tabulated by decade. All were listed up to the year 1900; random sampling of 100 theses per decade was used therafter. The information on marine theses comes from a poll I conducted while in the Office of Science and Technology; data

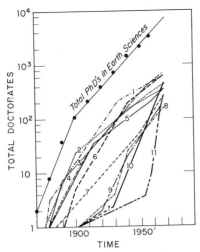

Fig. 7.5 Growth of doctorates in subfields plotted for comparison. Key: (1) Economic geology; (2) Igneous and metamorphic rocks; (3) Invertebrate paleontology; (4) Field geology; (5) Glacial geology-geomorphology; (6) Stratigraphy; (7) Geochemistry; (8) Micropaleontology; (9) Experimental; (10) Geophysics; (11) Oceonography.

is shown in Figure 7.4. The quality of the data for the first few degrees in each subfield varies with time because of the sampling technique. Every degree was counted until 1900, and therefore the shape of the growth curves is dependent only on the consistency of the classification into subfields. After 1900, however, statistical techniques were applied which could easily miss the first three or four papers. I have taken this into account in adjusting the first three degrees in geophysics in Figure 7.5, which shows all the growth curves plotted as a function of time. A remarkable pattern is evident. The older subfields follow exponential curves that are convex upward, indicating rapid initial growth that has gradually slowed. The doubling times for the first 10 degrees in these old subfields were only 3–4 years. For the next 90 degrees, doubling took about 10 years, and from 100 to 500 degrees it has been closer to 20 years. Moreover, the changes in doubling time themselves vary with time. Economic geology was the first subfield to have more than 1 degree, it doubled in 2 years, and then the growth slowed after only 3 de-

---

from 1958 through 1968 are from *Dissertation Abstracts,* University Microfilm. I assumed a linear variation during the eleven years.

grees. Invertebrate paleontology started to grow a little later, grew a little more slowly, and reached 7 degrees before the growth rate decelerated. The relationship is not perfect but almost so for the 5 oldest subfields. Stratigraphy, which began a little later, has a similar pattern. Geochemistry, on the other hand, had an uninterrupted exponential growth with a doubling time of about 11 years from the first degree in 1884 to the fifty-sixth in 1950. More recently it has accelerated with the flowering of isotopic analysis and radioactive dating.

The subfields younger than geochemistry follow exponential curves which are concave upward; after initially slow growth they accelerated. Moreover, the longer the initial period, the faster and longer lasting the spurt. In recent years, oceanography, the youngest subfield studied, has been doubling in two years or at the same spectacular rate as economic geology in the 1880's. The difference is that economic geology slowed after 3 degrees, and oceanography was last seen at 150 and climbing.

We may now speculate about the meaning of these growth curves and whether they are useful for forecasting future capacity and growth. The growth of doctorates in subfields is shown in Figure 7.6 in abstracted form along with the number of colleges and universities granting bachelor's and advanced degrees in the earth sciences.[15] The data can be explained in terms of: (1) rapid growth until all interested departments are staffed, and (2) a gradual increase in such departments. Economic geology thus produced three doctorates and thereby saturated the available professorships in 1880. Thereafter doctorates in this subfield grew at a rate adequate to staff new departments as they were formed and also to meet the needs of industry. Geomorphology doctorates began a little later and expanded a little more slowly, with the result that many more departments had openings for new staff, and this initial growth continued much longer than for economic geology. Geochemistry is an example of a subfield in which there was no intense interest and opportunities were few. Consequently, few departments saw any reason to appoint a geochemist, and there never was any rapid initial growth. On the other hand, the growth was sustained far longer than in the older

[15] Mathews and Little, "Geology and geography in the United States"; Lindquist, "Geology degrees during the decade of the fifties"; Downs and Henderson, "Student enrollment in earth science, 1967–68"; Chronic and Chronic, *Bibliography of theses*.

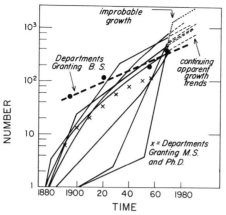

Fig. 7.6 Relationship between doctorates in subfields and the number of college departments in earth sciences.

fields, because many departments existed. Geophysics and oceanography are subfields that hardly existed before 1900 and that lacked tools for exploitation until fairly recently. When the means became available, they started to expand relatively rapidly, and this growth continued without interruption until the present, when the number of doctorates coincides with the number of available departments of earth science.

The explanation is logical in the sense that each new professor trains new students and this chain reaction continues as long as new professorships are available. The growth is inevitably slower thereafter, despite any urgent demands of research and industry. Assuming constant productivity, the fully staffed departments will produce a constant number of doctorates, which guarantees arithmetical rather than exponential growth.

This explanation has the corollary that the rapid expansion of a science depends on the invention of new subfields. Each available department is staffed in each new subfield, and then industry and research are filled, and then the growth of doctorates is locked to expansion of the population and economy and replacement of the aged. Other things being equal, far-sighted students should move into the new subfields that expand and add one more man to each department. This expansion seems typical but is not universal; vertebrate paleontology has had only 70 doctorates in almost a century.

So much for graduate education in geology. Is it similar in other sciences in that a significant fraction of the new science doctors have remained in universities as teachers and fed exponential growth? It appears so. In fact, the feedback in geology is relatively low.[16] It is more than twice as great in mathematics and social sciences and almost 50 percent greater in the life sciences and engineering.

Having observed something of the nature of the growth of graduate education and teaching departments, we can at last return to the question of the plant capacity of the system. We particularly wonder whether it is inadequate for the demand and alternately whether it is capable of an unprecedently rapid expansion. The most rapid expansion we have studied has been in oceanography with a doubling time of about 2 years for doctorates. In geophysics and in the rapidly expanding phases of other subfields, the doubling time is about 4–6 years; in older fields it is 10–15 years. We can compare these times with some models of expansion. The maximum plausible rate of expansion occurs if all new Ph.D.'s immediately become professors, each professor produces three Ph.D.'s per year, and it takes three years of graduate study to acquire a Ph.D. Such a chain reaction starting with one professor yields 487 doctorates in 12 years with a doubling time of only a year. A somewhat more likely model accepts the same conditions, except 2 out of each 3 new Ph.D.'s go into industry, research, or administration, and only 1 in 3 becomes a professor and produces more doctorates. This model gives only 39 Ph.D.'s in 12 years, but even so the doubling time is less than 2 years. We can conceive a third model in which it takes 6 years of graduate study for a degree, each professor takes on 3 new students per year, and only 1 Ph.D. in 3 becomes a professor. This gives a doubling time of about 2.5 years. It seems evident that these models are not impossible of realization, and thus a growth of a new subfield could be even faster than in oceanography. The actual development of oceanography has been achieved only partially in this mode. Much of the expansion of Ph.D. output has occurred by transfer of professors of physics and geology and biology into oceanography. This can yield extremely rapid expansion and initially amounts to little more than a change in the title of a thesis rather than in its content.

We may also consider a model capable of doubling as slowly as we

[16] R. H. Bolt, W. L. Koltun, and O. H. Levine, 1965, "Doctoral feedback into higher education," *Science 148*, 918–928.

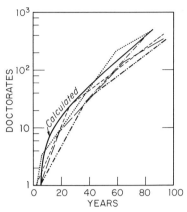

Fig. 7.7 Growth of doctorates in subfields in geology normalized to the same starting time and compared with a calculated model designed to produce a very slow increase.

observe in some subfields. The following one is capable of doubling in a period as long as 15 years:

1. Each professor takes on 1 new student a year.
2. It takes 6 years of graduate study to obtain a Ph.D.
3. Therefore, after an initial buildup, each professor has 6 students.
4. One student in 15 becomes a new professor.

This is hardly an exhausting program of graduate instruction, but it corresponds fairly well with the growth of the older subfields in the earth sciences (Fig. 7.7). We observed earlier that the number of professors and the output of graduates in geology are unrelated. We now find that in mature subfields the number of professors is controlled by the number of departments with positions, whereas the number of new doctorates is controlled by the slow expansion of the economy and population. Thus, there is hardly any relationship between the number of professors in a science and the number of degrees granted at any level. University professors in science are not primarily teachers.

The strength of this generalization for all science and engineering is readily seen in the ratio of doctoral degrees granted in 1961 to the doctorate-holding staff in universities.[17] Not all staff are professors, but even so the numbers are surprising (Table 7.1). Some 3,900

[17] Bolt et al., "Doctoral feedback into higher education."

TABLE 7.1. Number of Doctoral Degrees Granted per Doctorate-holding Staff Member in Science and Engineering in 1961 (r) for All Activities and Segments in Colleges and Universities, Tabulated by Academic Field

Academic Field	Doctoral Degrees Granted (1000's)	Doctorate-holding Staff of Scientists and Engineers (in full-time equivalents, in 1000's)	r
Mathematics	0.36	3.9	0.09
Physics, astronomy, and earth sciences	0.86	7.7	0.11
Chemistry	1.14	7.1	0.16
Engineering	1.01	4.8	0.21
All physical sciences, mathematics, and engineering	3.37	24.0	0.14
Biological sciences and agricultural and health sciences	1.80	14.0	0.13
Psychology	0.87	3.6	0.24
Social sciences	1.04	8.5	0.12
All life sciences	3.71	26.0	0.14
All fields of science and engineering	7.08	50.0	0.14

Source: R. H. Bolt, W. L. Koltun, and O. H. Levine, 1965, "Doctoral feedback into higher education," *Science 148*, 918–928.

mathematicians yielded 360 new ones, or less than 1 each per decade. The efficiency of output is little better in most other sciences. It seems that universities have a significant overcapacity for graduate training in sciences — or almost anything else — if professors ever become full-time teachers.

Without for a moment suggesting anything deliberate, it is interesting to note how many impediments to granting a doctorate have developed over the years. The basic requirements of the University of California, first written almost a century ago, do not include any particular graduate courses at all but, for some reason, most departments now require them. A knowledge of two languages is deemed essential. Make-up of undergraduate courses is common. Theses have to be in a certain format and, although it is hard to believe, on a certain type of paper. An outside observer might wonder about the objectives of the system. Radical students and conservative governors have proposed that the educational system ought to be capable of yielding more students. We now see that professional, as opposed to general education, is not designed to produce the maximum number of students but rather to try to match output to demand and thus prevent unemployment. If science professors spent all their time teaching and students were available to be taught, the reward for four to ten years of college study would be a relief check. It may already be the reward in some fields. Perhaps this explains the growing feeling that a Doctor of Science degree, after the Ph.D., will soon be required for a qualified scientist. An extra five years of training would delay output, and mortality would even reduce it.

THE ROLE OF THE UNIVERSITY SCIENCE PROFESSOR

We now have many indications of what professors of science in universities do not do. Are they idle? The University of California made a specal study of this question and concluded, as I recall, that the average professor worked 60 hours a week on university business. Are they incompetent? The Peter Principle is that each worker advances to a level at which he is incompetent, and thus we would expect it.[18] However, professors of science in great universities in this country generally are competent.

The Peter Principle is almost universally applicable but not quite. It works only under one or both of the following circumstances: (1) competence for a particular job cannot be determined before hiring or even before the worker achieves permanent status, or (2) a worker competent at one job has the opportunity for advancement to another at which he is not known to be competent — which returns him to stage (1).

It is awe-inspiring to see how universities generally have prevented these circumstances from arising. Professors occasionally become deans and chancellors, and as such they clearly may be as incompetent as the next man subject to the Peter Principle. However, until the development of the multiversity with its conventional bureaucracy, very few academic administrators were needed. Even so, we quickly grant that all administrators rise to their level of incompetence and note that universities suppress the urge to administer

[18] L. F. Peter and R. Hull, 1969, *The Peter principle* (New York, William Morrow & Co., Inc.).

by simply not paying any extra salary to department chairmen. Instead, the job is rotated among nonadministrators. It is only the working professors who appear to be competent. This statement might appear ludicrous considering how many patently incompetent teachers exist in even the greatest universities. It is true nonetheless, because the competence of university professors of science, and perhaps all university professors, lies in research rather than in teaching. This fact, which appears to have been forgotten in some recent discussions about teaching, is readily apparent from even the most cursory inspection of their role.

Presumably the great universities with the first chairs in science wanted to hire professors who would be skillful at teaching the subject to students. These would be imaginative, dedicated scholars, gifted in communicating ideas, and sensitive to the needs of students. On the record, it appears that it was immediately recognized that these characteristics were impossible to identify before hiring or before a teacher was around long enough to achieve tenure. Thus, if professors were to be hired on their putative abilities as teachers, the whole system of university education in the sciences would have fallen victim of the Peter Principle. Every professor would have been incompetent for most of his career. The naive idea of educating the young by hiring good teachers thus was abandoned by universities of the first class.

Although it was virtually impossible to identify even a good teacher before hiring him, it was quite simple to identify a superb student by conventional tests. Thus, the great universities solved the teaching problem by admitting only such students as would be capable of learning from any teacher no matter how incompetent. This was achieved by luring student applicants with munificent fellowships, unparalleled facilities of libraries and laboratories, the opportunity to hobnob with a group of famous professors and to have as classmates those who would become famous, and ultimately the prospect of a prestigious degree. If the bait was adequate, applications exceeded openings, and only the best students were admitted. The success of this approach is demonstrated by the fact that the science graduates of such places as Chicago and M.I.T. consistently score near the top in graduate record examinations and readily obtain doctorates from other great universities. In general, they are no less able when they finish college than when they start. Success is also demonstrated by the level of affluence of alumni, who may be expected to reinforce endowment and thereby make the fellowships even more desirable.

In this university system of education the primary mission of the professor is to become famous and thereby lure the best students. An earlier study of this question indicates that professors are hired primarily according to their estimated ability to attract the best colleagues.[19] This apparent discrepancy may be attributed to the inclusion of humanities professors in the sample. Therefore a more general conclusion is that, although all professors are expected to attract rather than teach, some attract students and some each other. The secondary mission of the science professor is to illustrate by his life style how someone behaves who is competent at something. The problem of the university thus is reduced to merely finding people capable of carrying out these missions. The criterion used is skill in scientific research which is relatively easy to measure and even predict. First, the candidate must show that he is a good student, because it is at least helpful if the professors are mostly smarter than most of the students. Then he demonstrates that he can do research and obtains a doctorate. Then he serves an apprenticeship as lecturer or assistant professor. For some years after hiring he has only a yearly contract which need not be renewed. If his research does not seem to be prospering, he is dismissed and becomes a professor with tenure at some lesser college.[20] Thus, there is no commitment to the Peter Principle until he becomes an associate professor, by which time he is known to be good at research in something. Above him lies only one promotion, to full professor, and this merely gives him more pay to do the same thing. This is true of endowed chairs and super-scale professorships, which are rewards rather than different jobs. In sum, a professor achieves the mission for which he was hired, namely, doing research and attaining scientific prestige. If he becomes famous and lures many students, he receives additional rewards.

It is manifest that this system will not work unless the students are carefully selected. There is no prospect that the teaching of science in state colleges and junior colleges can function as it does in great universities. The mass of students are taught by teachers who are not

[19] T. Caplow and R. J. McGee, 1961, *The academic marketplace* (New York, Science Editions, Inc.).
[20] Caplow and McGee, *The academic marketplace*.

skilled in research. Mercifully, some are skilled in teaching, but only the small fraction who beat the Peter Principle for a while. Interestingly enough, these nonresearch colleges provide, for the first time, a proving ground for teachers. If the universities want demonstrably able teachers for undergraduates, they can get them from the colleges. It is by no means clear that this would be a good idea. The best teachers would be stripped from the colleges leaving the average students in less competent hands. These are the students most likely to need the best teachers. Meanwhile the best students, who have less need for good teachers, would get them. Even the universities might not benefit much. A research scientist who becomes incompetent leaves his publications and prestige behind so the university has something to draw students. This effect is observable when students register to take courses from distinguished professors who are dead, senile, or departed for another campus. What would the university get from a teacher who becomes incompetent?

EDUCATION IN A STEADY-STATE SCIENCE

Just as there are differences in research in fields growing at different rates, so we may expect differences in education. Geology provides excellent though limited data on the character of some of the differences because it was a normal science in the nineteenth century and one with steady-state growth during much of the present century.

The curriculum in the early years cannot have been very different from that in other sciences, because most of the specialties which now exist had not been invented. The geology major thus was educated in physics, chemistry, mathematics, biology, and English, perhaps as much as in geology. This is evident in the senior class examination in geology given by Professor Joseph LeConte in November 1857 at the University of South Carolina.

1. What is the law of variation of the carrying power of running water? and prove the law.
2. Explain why bars are formed at the mouths of rivers.
3. Explain why sedimentary rocks are always stratified and conversely.
4. Under what circumstances are natural levees formed, and why?
5. Give a succinct account of Forbes' Theory of Glacier motion, and the proofs on the truth of this theory.
6. To find the depth of an earthquake shock when the velocity of the

Education

spherical wave, the velocity of the surface wave, and the distance from the center of principal surface disturbance was given. Case 2d on a spherical surface.

7. What are the five conditions which limit the growth of corals?
8. What are the three kinds of coral reefs, and briefly characterize each.
9. Explain the manner in which *islands* are formed upon coral reefs.
10. Give an account of Tyndall's Theory of Slaty Cleavage.
11. Give the proofs that coal is of vegetable origin.[21]

Most of the questions are concerned with physical laws that apply to geology. In another contrast with the present, a student could pass the senior examination with hardly any knowledge of geological jargon. There are other types of evidence that the early geologists were broadly educated. The first volume of the *Bulletin of the Geological Society,* for example, contains an equation involving integral calculus in a paper about the geomorphology of Alaska.[22] It is an interesting exercise to see how many geologists trained during the steady-state period ever had enough calculus to solve such an equation even when they were students. Some data are provided by the Curriculum and Standards Committee of the National Association of Geology Teachers which polled 208 departments regarding the mathematics requirements for a degree in geology.[23] Minimum standards are as follows: algebra, 4 percent; trigonometry, 21 percent; analytical geometry, 5 percent; "one year" — presumably pre-calculus, 15 percent. In sum, almost half the graduates were not required to understand the mathematics in a paper on field work written 70 years earlier.

Gradually subfields developed, and education became more specialized. During the steady-state period it was quite possible to obtain an advanced degree in geology by merely learning a great deal about stratigraphy, paleontology, structural geology, geomorphology, mineralogy, petrology, and the other subdivisions of the field. Indeed, even a knowledge of all aspects of geology was not universally viewed as necessary. At Harvard in the late 1940's a paleontologist was not required to take petrology to obtain a bachelor's degree. It seems unlikely that the requirement broadened in graduate school.

[21] *Geotimes 12* (1967), no. 4, 18. Discovered by L. L. Smith, emeritus chairman of the same department.
[22] I. C. Russell, 1890, "Notes on the surface geology of Alaska," *Geological Society of America Bulletin 1,* 99–162.
[23] C. E. Prouty, 1961, "Curriculum survey," *Geotimes 6,* no. 3, 28.

Meanwhile a different concept of geological education began to evolve with the development of geophysics. Once again a student was required to have a broad education in basic sciences. The limitations of time dictated a corresponding decrease in special courses in geology. In most places these different concepts were accommodated by establishing a major in geophysics in addition to one in geology. At a few younger colleges, such as Caltech, the geology departments were founded to follow the basic science concept. Consequently, all students were exposed to several years of college level science and mathematics before the geologists began to take courses in geology. The geophysicists in such departments took far more physics and mathematics than geology as undergraduates.

This dichotomy continues. Through the usual processes of historical change the concept of specialized education which was radical a century ago is now defended by conservatives against the attacks of radical advocates of general scientific training. Committees have been established within the geological societies to design the ideal curriculum. Individual departments are rifted by divergent views among the faculty.

From this brief analysis it appears that education in a steady-state field may tend to become ever narrower. Most professors are more broadly trained than their predecessors, but those in a steady-state field are the reverse. The professors become increasingly isolated and their students even more so. It is little wonder that the field falls farther and farther behind until it may become incapable of bridging the embayments in the research front.

It appears that any subfield that persists long enough may gradually grow more slowly. This is suggested by a comparison of growth rates for papers and doctorates in given subfields (Fig. 7.4). The magnitude of the subfields is not known, only the rate of growth. Thus, the growth in different subfields can be compared only by normalization. Assume, purely for this purpose, that the same number of papers were published in each subfield of interest in 1968. Figure 7.8, which is based on this assumption, shows that the growth of subfields has the same general character as that of doctorates. The older ones such as geomorphology are growing relatively slowly, and the younger ones such as geophysics are growing rapidly. The sole exception is economic geology, in which papers are expanding rapidly but doctorates are not.

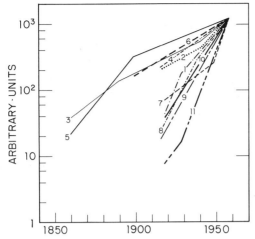

Fig. 7.8 Growth of subfields plotted as if total of papers in each were the same at present. By extrapolation it appears that the earlier a subfield began to grow, the slower it is growing now.

Three possible basic relationships exist between growth of doctorates and literature in a subfield. They may be at the same rate or diverge or converge. More complicated patterns may result from the combination of these basic possibilities. Economic geology, as we have just noted, shows a divergent pattern. This means that in some way the ratio of papers to doctorates is increasing. Either the doctors are becoming more active (efficient?) in research or nondoctors are writing more research papers.

Economic geology is the only subfield with a consistent divergent pattern, and only the "experimental" subfield has a long-term parallel pattern. The characteristic pattern is convergence although for at least a decade some have been divergent. By all other criteria we have established, this is a highly mixed bag clearly unrelated to growth rates.

In general, the outputs of doctorates and of research papers tend to converge in subfields of geology regardless of how rapidly they are growing. This can be expressed as a decrease in the research productivity of all Ph.D.'s in a subfield as it matures. We know that professors are not producing more graduates nor doctorates as a field matures. We know that more and more new men go into occupations other than teaching as it matures. With constant efficiency, we would expect a marked increase in average productivity of research instead of the reverse. What is the explanation? Is research not followed by publications? Possibly this reflects the same stifling of research which prevailed in the earth sciences in the first half of the century. As the subfield matures, more and more effort is devoted to bibliographic compilation, for example, and less to measurements in the laboratory.

It is discouraging to observe the convergence in young and vigorous subfields regardless of the explanation. On the other hand, there are grounds for hope. For some reason, some subfields now show a divergent pattern. Perhaps they are new subfields launched from and superimposed on older ones. Thus, isotope geochemistry is very different from the older geochemistry; high pressure experiments have revitalized igneous petrology; and paleoecology has brought a new focus to invertebrate paleontology. If this is correct, it will only be after the new subfields become distinct that they can be tested for convergence or continuing divergence.

STATE OF THE SCIENCE EDUCATION INDUSTRY

American education in the sciences is most widely acclaimed. At the moment, however, the industry seems to be suffering from excess capacity and inventory build-up, and the output seems to exceed demand. This may be temporary and in itself not much cause for concern about long-term growth. Output has fluctuated in response to political and economic pressures in the past and has always recovered. The underlying reason for concern is that the industry is geared to rapid growth with a substantial allocation of output back into expansion.

The long-sustained average growth of output has been much faster than that of the population. This means a larger proportion of the scientifically educable fraction of the population is receiving education in sciences. A limit to this fraction clearly exists and perhaps is being approached. If so, output of the present type cannot grow faster than the population. Thus, the best hope for those who want to continue to expand the industry is to invent another type of output of which recycling, or retrofitting, appears the most promising. If every scientist were given a new education every fifteen years most of them would be much more useful. Moreover, the in-

Education

dustry of educating scientists could continue to expand for a few more doubling periods.

So much for the prospects for the industry. What of the prospects for the students who are its output? I have spoken to students on several campuses and in many departments ranging through the arts and sciences. I have yet to find one who had ever been told anything much by his professors about his prospects for a job, or one who was not interested in the subject once it was raised. It is time for professors to take more responsibility in investigating what will happen to students. It has been assumed in many leading science departments that a professor helps place the people who do doctoral research under him. This worked very well during the period of general expansion of education, but is not now so successful. Many students were never afforded such support at the most critical stage in their careers. All of them need and deserve advice and help.

In contrast to the industry, the long-term prospects for the employment of scientists presumably are very good if they enter expanding fields or seek jobs outside of universities. Training to become a scientist is hard compared to other forms of education; and afterward the hours are long and the required dedication is great. Thus, I cannot recommend it to anyone who expects to do research as a form of work. Do something else if you want to make money. Find another field if you want stability, security, or order.

> To follow knowledge, like a sinking star,
> Beyond the utmost bound of human thought.[24]

That is the only reason to be a scientist. With that motivation science is not work at all, and the hours seem long when you are not doing it.

Nevertheless, the world intrudes on nirvana. Given the basic attitude suitable for a scientist, a student might want to smooth his path by deliberately seeking a promising field. Some will not. Ornithologists, who seem to be princes one and all, are often dedicated from the first sighting of a bird sanctuary. Most students, however, hardly know whether they want to be scientists let alone biologists or molecular biologists. For them the early chapters of this book may be of use. Having identified a field that sounds interesting, seek

[24] Alfred, Lord Tennyson, "Ulysses."

out the principal journals in the library. Not just general journals, but the ones which contain research papers in that very field. A more advanced student can look in the indexes in *Nature* and *Science* if in doubt. With the pertinent journals in hand, analyze the distribution of citation ages according to the procedure in Chapter 2. That will give the growth rate and provide the basis for estimating career possibilities. Ask the professors in the field for their evaluation, but make sure that they have gone through the same procedure or have made some other kind of study. Clamor for courses in the history and sociology of science and for introductory material in each advanced science course. It will help professors as well as students to learn about supply and demand for their services.

8 A Department of Science

Broadly speaking, the federal government consists of many different types of organizations of interest to scientists. At the top, the Congress has had little contact with science except in legislative hearings. On the other hand, the White House has both a staff (the Office of Science and Technology) and an elaborate advisory network centered on the President's Science Advisory Committee. This dichotomy may tend to put scientists in a hostile position in the eyes of the Congress and certainly inhibits an easy exchange of information. More balanced availability of advice is desirable.

Most government interest in science is not at the top, but in the operating agencies, which are of three types which we can call dodgers, doers, and dispensers. The first type are those charged with the regulation of major industries but which lack both the means and the will to do so. They are readily identified as being under investigation by Ralph Nader. They appear to spend more on public relations presentations than on science and are of little consequence in this analysis.

The second group of agencies are those which do things related to science, such as monitor and predict the weather or map potential mineral resources. These can also be readily identified by dividing the annual budget of the agency by the total number of employees. The resulting ratio should be about equal to the annual salary of a middle level civil servant. In 1964, the ratio was between $10,000 and $15,000 per person for six government agencies of this type. They include such apparently different organizations as the Weather Bureau, Naval Oceanographic Office, Bureau of Mines, Coast and Geodetic Survey, Army Map Service, and the Geological Survey. This group is of great importance to scientists because it employs so many of them, and we shall analyze it in detail. It is likewise important to the nation because of the utility of its product. However, it is fairly obvious that such agencies spend all their appropriations on salaries for themselves, and consequently they are of little interest to anyone else looking out for himself.

The federal government has always been engaged in scientific activities and largely in this group of agencies until recently. By 1884, the National Academy of Sciences recommended that many of the already numerous agencies be combined into a Department of Science.[1] Essentially the same recommendation has been made from time to time ever since and always engages some fraction of the interest of the Congress or the Administration.[2] These organizational changes would be highly desirable for the nation, which is badly in need of environmental services. Because of the nature of the agencies, however, they appear an inappropriate medium for a Department of Science — unless they metamorphose into dispensers.

The third class of agencies of interest to scientists are those dispensing money to people who are not employees. They are also readily identified by the value of the dollars to people ratio. In such agencies as the Atomic Energy Commission, NASA, and the National Science Foundation, the ratio in 1964 was $130,000 to $400,000 per employee. These are the kinds of agencies that interest people on the outside whether Congressmen, contractors, or chemists. They have money to dispense.

ENVIRONMENTAL SCIENCES IN GOVERNMENT

Responsibility for and research on the environmental sciences are divided among many federal and civilian agencies in the United States and indeed in most nations. They are created within, and assigned to, government departments in quite an arbitrary manner. For example, the Bureau of Mines, now in Interior, originally budded off from the Geological Survey in Interior, but from 1925 to 1934 it was in the Department of Commerce.

The environmental sciences are also of urgent concern to the

[1] J. W. Powell, 1885, *On the organization of scientific work of the general government* (Washington, Government Printing Office).

[2] Office of Science and Technology, 1966, *Effective use of the sea* (Washington, Government Printing Office); National Academy of Sciences, 1970, *Institutions for effective management of the environment* (Washington, National Academy of Sciences).

A Department of Science

Army, Navy, and Air Force, which need information for operations in places outside the United States. Consequently, the Navy has an Oceanographic Office, formerly the Hydrographic Office, which collects many of the same types of information as the Coast and Geodetic Survey. The Army has the Army Map Service, which does about the same things as some divisions of the Geological Survey. The Air Force operates a global weather service matching the Weather Bureau. All the military agencies are roughly the same size as the civilian ones. It is apparent that the nation is well endowed with government agencies concerned with the environment. Occasionally suspicions have arisen in the Congress and the White House that there might be just a little duplication of effort, and a call has gone out for economy and reorganization. One such reorganization was proposed in 1969 by the National Commission on Marine Science, Engineering and Resources.[3] Others have been studied almost without surcease in recent years in the Executive Office of the President. However, the problem is not a new one. John Wesley Powell, Director of the Geological Survey, testified on the matter in 1884 before the Joint Commission of the Congress appointed to "consider the present organizations of the Signal Service, Geological Survey, Coast and Geodetic Survey, and the Hydrographic Office of the Navy Department, with a view to secure greater efficiency and economy of administration . . ."[4] He demonstrated duplication and inefficiency; he proposed total reorganization. Nothing came of it, and little has happened in numerous hearings since.

GROWTH OF ENVIRONMENTAL AGENCIES

Figure 8.1 shows the annual budgets at 5-year intervals of civilian and military environmental agencies, the civilian component of the whole budget, and the whole budget. It is clear that each component and the whole generally grow exponentially. The entire federal budget was $40 million in 1850 and about $100,000 million in 1965, with fairly constant growth except during wars. Barring wartime peaks, the expense of the federal government has increased

---

[3] National Commission on Marine Science, Engineering and Resources, 1969, *Our nation and the sea* (Washington, Government Printing Office).
[4] J. W. Powell, *On the organization of scientific work.*

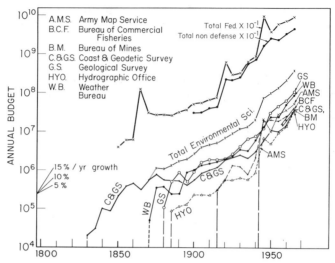

Fig. 8.1 Annual budgets of federal agencies concerned with the environmental sciences compared with the nondefense and total federal budget.

every year since 1885 and all but once since 1850. Since 1900, when separate data begin, the civilian component of the federal budget has likewise grown exponentially with a doubling time of 9 years. If we turn to the sum of the budgets of all the environmental agencies, we see the very striking fact that it has grown exponentially with exactly the same doubling time of 9 years since the founding of the Coast and Geodetic Survey in 1832. In 1965 the civilian total was $47 billion and the environmental total was $0.4 billion. Inasmuch as the ratio has remained constant, the government has spent and spends about 1 percent of its funds to find out about the weather, map the land, and generally keep track of the environment. This fraction of the budget has been adequate to see the mineral resources depleted, the soil ravaged, and the air and water polluted. Presumably it will cost a considerably larger fraction to maintain the environment in its present sorry state. Certainly it will cost much more to restore it.

The total environmental budget has remained 1 percent, while the number of agencies has increased. Thus the average fraction of the budget for each agency has steadily decreased. The Coast and Geodetic Survey was still the only environmental agency in 1865. By 1900 there were three additional agencies, and the Coast and Geodetic Survey received only 20 percent of the environmental

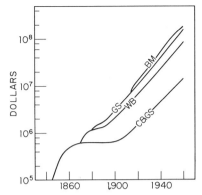

Fig. 8.2 During the nineteenth century each new agency was superimposed on a relatively stable older one.

budget. This dropped to 10 percent by 1965, or slightly less than its fair share of 14 percent as one of the seven agencies.

The approximate long-term growth appears in Figure 8.2, in which the annual budgets of four civilian agencies are summed. The sustained growth of the environmental component of government was achieved in the last half of the nineteenth century by holding the budgets of the older agencies relatively constant and building up the new ones.

The regular growth suggests that these agencies have a life of their own that is generally independent of forces outside of government. This view is reinforced by the remarkable regularity of the growth patterns of the civilian agencies. The four with a significantly long history are the Geological Survey, Weather Bureau, Coast and Geodetic Survey, and Bureau of Mines. They were founded over a period of 85 years and apparently randomly with regard to wars or other major events affecting funding. In Figure 8.3 the growth of the budgets is plotted as though all the agencies began at the same time and with the same budget. A striking pattern emerges for the first 50 years of growth. During the first 10 years the funding increases five- to eightfold. During the next 5 to 10 years it decreases. In 1840–1845 the Coast Survey budget decreased to only 90 percent of the peak, but the successively younger agencies were cut to 59 percent, 60 percent, and 68 percent of their peaks. After this period of decline, each agency prospered, and 27 years after founding all had grown the same amount, namely, twentyfold. The doubling period was only 6 years even including the intermediate decline. By age

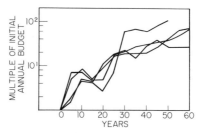

Fig. 8.3 Growth in the annual budgets of federal environmental agencies normalized to the same founding date and the same starting budget.

30, the agencies were all mature and growing at very slow rates with doubling times of 20 to 40 years.

Allow me to propose a scenario compatible with these orderly patterns of growth. Periodically the Congress and the Executive identify some component of the environment which needs study or monitoring in connection with the aims of the nation. The Coast and Geodetic Survey was established to help prevent ships from running aground. Inevitably the formation of a new agency requires hearings and debate, and nothing happens until the need has grown acute. Consequently, once the agency is formed it grows rapidly for a decade. Having achieved momentum, the growth exceeds the need or else outruns the supply of technical staff, and the agency languishes for 5 or 10 years before growth renews. The further history of the agency is sustained growth but at a rate slower than government in general and thus with a high probability of increasing stagnation.

As the nation and government grow, new environmental problems arise. The response of the Congress and the Executive is not to expand the old agencies into a new field. The old agencies show signs of age and are viewed as burgeoning bureaucracies. Instead, a new agency is formed which then grows vigorously and repeats the pattern. Taking the intermediate view, this seems a very reasonable way to deal with a problem. Difficulties arise only if, in the long run, the problem is solved because then the agency has nothing to do. Unfortunately most environmental problems seem to persist or worsen.

Meanwhile, it may be interesting to speculate about what would happen if the government ever followed the course often considered, namely, the solution of a new problem by consolidation or

A Department of Science

expansion of existing agencies. We visualize a mature agency established to deal with problem X. A related problem, Z, is identified. This agency would begin with a large administration attuned to the importance of X but not Z. It would begin with all higher-level technical positions filled by men educated in X some 20 to 40 years before. It would begin with fully operational bowling leagues, bridge clubs, and group travel associations and a staff with large quantities of accrued annual leave. Would it accomplish the solution of problem Z as rapidly and cheaply as a new agency? We emerge from these speculations with enhanced appreciation for the wisdom of our elected representatives who have always adopted a different solution.

The sensible organization to solve a new problem appears to be a new agency. What of the organization to deal with such long persistent problems of the weather and the grounding of ships? For example, is it worthwhile to combine two old agencies responsible for the old problems X and Y? It is hard to see how anything would be lost and some economies would result. Only very few top officials need be added to control the combined bureaucracies.

The environmental scientists continue to measure and predict X and Y, activities in which they are trained and experienced. Consequently, the present disadvantages are minimal. The future advantages are many and substantial. Perhaps most important is the enhanced capability to do modern data-processing. The Weather Bureau has enormous computers, for example, which can be used to handle all manner of environmental data. Big computers generally are much more efficient than small ones, and they can be used to offer more environmental information in less time. The many other advantages are discussed in the reports on government hearings on reorganization and need not be repeated here. It appears that they outweigh the disadvantages and that the old stagnating environmental agencies should be combined. Perhaps they would take on a new vigor and begin to obtain at least a constant fraction of the federal budget.

DEPARTMENT OF SCIENCE AND TECHNOLOGY

At present, science and technology are viewed in Washington as things that are used. This is hardly surprising considering how rapidly they have expanded and considering the advanced average

age of the officials who are in power. It is generally stated, at least in the Bureau of the Budget, that there is no need for anything like a Department of Science because the government has no science mission. Instead, the Department of Transportation has a need for scientific support to meet its assigned mission to strengthen and support transportation for the good of the nation. Likewise, the other departments and agencies have missions and use science and technology. However, departments and missions change. Indeed, most of the larger agencies now existing are quite new, and before them the Department of Labor and the Department of the Air Force were not so very old.

There is a well-established precedent for forming new departments provided a sufficiently important and homogeneous mission can be identified. Whether a Department of Science is needed depends entirely on how we look at things. We have a Department of Agriculture and one of Transportation, and both are doubtless needed. Not only do they help within the counry, but the groups they service are among the principal sources of foreign exchange. Let us view these departments in a somewhat different way. Our agriculture is the most efficient in the world. Is it because we have lots of farmers using primitive tools to cultivate primitive grain stocks? Or is it because we have elaborate machines, chemical fertilizers, chemical and now biological pest control, hybrid corn, and special strains of fruits and vegetables that can be picked by machine?

What of transportation? The low-technology services such as trains fall into disuse, and high-technology airplanes become ever more popular. We export jet planes and the advanced technology to make them work.

Granting that the organization of the government is arbitrary, a good case can be made for a Department of Science and Technology with the mission of advancing a broad range of activities from basic research through development to testing. In short, it would provide encouragement and support for what is becoming one of the most important concerns of this nation. A case can be made, but who will support it with determination and logic and muscle? Not the unorganized scientists who come hat in hand to Washington, not the engineers and technicians who will be increasingly unemployed; a General Union of Technologists and Scientists could.

A Department of Science

The environmental departments of the federal government do not appear to be a logical base upon which to construct a Department of Science and Technology, because, on the record, they are losers. They are important to the government and the nation and to the scientists who work for them but, judging by the support they receive, nobody loves them.

This can change, or at least it did once when the Department of Agriculture was transformed from a highly successful doer to a highly successful dispenser early in the century. The research agency of 1875 grew rapidly to 1895, caught its breath for 5 years, and then doubled its budget in the next 5 to about $4 million. In 1910 it was almost $12 million and still doing research. In 1915 it was $20 million, but spending a quarter of that for meat inspection and buying forest lands. In 1920 the transformation was complete. The budget was $196 million, of which $145 million was for rural road construction. Now it is about $8,000 million, most of it for farm subsidies. It serves a rapidly decreasing constituency of farmers, many or most of whom farm only part time.

That could happen to an environmental agency, but it seems unlikely. The Department of Science and Technology thus should be based on successful dispensing agencies such as the AEC, NASA, and the National Science Foundation. "Successful" may seem ironical, considering what is happening to the budgets of these agencies. They are, nevertheless, successful at what they do. The problem is that NASA patently has an outmoded mission, and the NSF has no organized constituency. A Department of Science and Technology would long since have shifted some of the space scientists to urban problems or pollution after retraining as needed. As it is now, the departments responsible for housing, urban affairs, and transportation have an urgent need for technological support, but no way to fund it and perhaps not much urge to. A department responsible for technological matters would feel otherwise. As to the lack of an organized constituency favoring a Department of Science, we shall now turn to the possibility — indeed the necessity — of forming one.

9 Scientists in Society

Scientists and engineers have had easy lives for a long, long time — a fitting reward for ability and the discipline and youthful sacrifices to technical schooling. Expansion was everywhere, students were readily absorbed into the work force, jobs were abundant, and all was right with the world. The research scientist was a wizard: wise, disinterested in worldly affairs, and pursuing strange dreams for the welfare of man. This image persisted while scientists and engineers multiplied a thousandfold and grew affluent on the public purse or in those industries directly nourished by that purse. It has persisted, but undergone a sickening change. The scientist is a warlock responsible for the terrible weapons that threaten life on earth; the engineer and the technologist are the ones who forged them. Their dreams are evil dreams; their gods are devils; and their concerns are inimical to humanity. As the image has decayed, so has the easy life. The particle physicist and the aerospace engineer are unemployed, and others will follow.

The superior, disinterested image may have been appropriate, although smug, in its time, but it seems a little dated when scientists alone are about as numerous as teamsters, and combined with engineers, technicians, and other natural allies they are as common as full-time farmers. The image was all right, although unbecoming, as long as it worked, but it began to be eroded some time ago. Isolation begets fear in others, and to it scientists owe some of their image among the young as creatures carrying out the will of evil masters. An attitude of superiority begets envy, and who speaks up for science when it is under attack?

It is time for scientists to start thinking of themselves as workers. An elitist attitude may have been difficult to avoid when a small number of men were respected and a few were admired. It was ample compensation for long hours and low pay. "Give me enough ribbon," said Napoleon, "and I can conquer the world." However, when many people devote long hours to science and are unhonored and unsung, they are exploited workers rather than an elite.

When scientists learn to adjust their own image of themselves, they will be ready to do what other workers of the world have done — unite. They can go either into a craft union, perhaps "guild" is a better name, or into an industrial union. A Guild of Scientists would attempt to bring prosperity and security to scientists, who would be narrowly defined. A General Union of Technologists and Scientists would take the bolder view that maximum strength lies in uniting the largest possible group of interdependent workers. Scientists, engineers, and technicians are but a few of the appropriate groups that would belong to the general union.

The old order has changed, and the question is what will take its place, not whether something will. Scientists who will make no attempt to control their destinies deserve what will happen to them. I speak not of the prize-winning scientists who are quite secure — unless the disruption of the meeting of the National Academy of Sciences at Dartmouth by activists protesting the uses of research is a forerunner of things to come. I am principally concerned with those who do not win prizes, who are equally dedicated and hard working, but who may have a bleak future if present trends continue. I am also concerned, as they are, with the effects that will follow if scientific research continues to be discredited as it is beginning to be in this country.

The development of the scientific method and the construction of the edifice of science are the greatest group achievements of mankind. Art, music, and poetry may involve greater achievements, but they are individual efforts and should be compared with the discovery of Newton's laws or of Einstein's relativity theory. In contrast, most human group activities, which include jihads as well as justice, are remarkably ineffective and show little sign of change for the better. Science is the only group activity that seems capable, at present, of indefinite improvement and advancement, because it builds on a provable base. We already know much about nature and a little about living with it, but unfortunately we know hardly anything about controlling ourselves. The achievements of science are used to escalate our ability to fight wars, ravage the earth, fill the world with people, and generally spotlight the failures of politics and government. We are no better at controlling the hydrogen bomb than the crossbow. Is it any wonder that antiscience forces are emerging? They may realize that the fault does not lie with science, but attacking it is one way to try to take lightning out of the hands of politicians who have always been burned by matches.

The antiscience groundswell is as wrong as it can be. We are utterly committed to a technological civilization for the foreseeable future, whether we like it or not. Without it we cannot support the existing population, let alone any increase. The alternative to technology is starvation, disease, and mega-death, followed, judging by the human record, by war with the crossbow. The problem is how to live with technology, because we cannot as a group, or even individually for long, live without it. We have groups in New Mexico, as we had in Ohio a century ago, that attempt to withdraw from technology and live naturally. I respect their motives and think it might be possible after a few generations of birth control. However, those of them who scavenge on society, fly to India to stock up on hash, play radios or hi-fis, or plant hybrid corn are hardly living without technology.

The problem is how to bring technology under control, and this will be impossible to solve even with major social and economic changes. The best that such changes can accomplish alone is to bring reason into the use of present technology. All the politicians in the world, acting rationally, cannot put a technological world back together again. They can abolish oil and thereby eliminate smog, but at present, without transportation driven by oil, we would starve. The smog problem and the transportation problem in general, including the extreme of making people stay home, can only be solved by advances in technology, and they require scientific research and engineering development.

We speak of preserving the environment or even restoring its quality, but how can we, if we know hardly anything about it? Who will tell us? It is highly unlikely that we have even identified many of the problems that will beset us. What will happen when the first blight hits the new strains of rice and wheat that feed the open mouths of Asia and South America? Who will invent the means of completely reprocessing sewage and solid waste? Politicians? Financiers? Everyone can make some kind of contribution if he wishes, and indeed pollution will not be abated without politicians and financiers, but even with their help the problems will require scientific research.

The prosperity of the scientific enterprise is vital to everyone. So is the prosperity of scientists, at least to the extent that they are needed for the work. Scientists themselves naturally have a broader interest in their own prosperity and prestige and self-esteem, and it is to the question of how this self-interest can be furthered that we now turn.

THE GUILD OF SCIENTISTS

How will scientists react if the growth of science slackens and there is no excess of demand in any field? Their behavior in matters of employment up to now has been positively saintly. Why should it be otherwise for workers in a perpetual labor shortage? It should not be assumed that their attitudes will remain the same in a labor surplus. Until recently they have had the expectable benefits of a trade union without the effort of organization or payment of dues. Demand beyond supply provides all the necessary aid and protection. Now it may be time to form an organization, a guild, which the dictionary defines as "an association of men with kindred pursuits or common interests or aims for mutual aid and protection."

It may seem improbable that scientists will ever acknowledge that they are workers and be willing to be organized; but a little unemployment seems to be remarkably effective in sweeping away prejudices. The general attitude of otherwise similar scientists in three different employment markets is admirably illuminated in polls conducted by the members of the Committee on Professional Employment Standards of the American Institute of Professional Geologists.[1] The first group polled, 121 engineering geologists, seemed happy; 82 percent felt reasonably secure in their jobs, and 71 percent would again choose to be geologists if they could start over. The engineering and construction companies for which they work are generally expanding and have the prospects for continuing to do so.

Employees have an overwhelming desire to see their ideas or work put to use and to have the opportunity to be creative. Earnings, secu-

[1] E. L. Krinitzsky and M. T. Rader, Jr., 1969, "Problems in employment," *Geotimes* *14*, no. 2, 16–19.

rity, and professional status rank low in comparison. In fact persons under 30 rated creativity and work put to use as 100 per cent of a job's meaning . . . Status and security are more desired after 45 but are never of predominating importance.[2]

The second group, a total of 456 oil geologists, also said that they were principally motivated by a desire to be creative and see their ideas used, although fully a quarter ranked earnings as of greatest importance. However, these people were working in quite different circumstances from the engineering geologists. Exploration for oil is not expanding like engineering geology. Twice within the memories of many of them their companies had reduced the staff of geologists. Consequently, only 75 percent felt reasonably secure, and no more than 54 percent, barely half, would be geologists if they had another chance. Under the circumstances, the stated motivation of these workers seems strange. Why do they not consider themselves as workers in need of the mutual aid and protection provided by a guild? Their view of their helplessness is reflected in the answer to the question about why individual geologists are fired. A third of them listed "technical incompetence," but almost two-thirds cited either company operations that became uneconomical or reorganization by management. Nor do they have much faith that seniority provides security. Quite the contrary: many respondents observed that older men are fired and replaced by younger ones and that younger men are jumped into higher positions. Are these career risks balanced by commensurate benefits such as high pay? The respondents think not.

Perhaps the geologists still remaining with oil companies retain some optimism because they are at least better off than those who have retired, resigned, or been fired. These form the third group polled. Those leaving oil companies can also leave geology and stay retired or seek employment in some nongeological job at much lower pay. However, many elect to practice as geological consultants, and, in fact, the 364 responding consultants mainly resigned voluntarily, or say they did, and presumably are relatively well off compared to those who had no choice and were fired. This group estimates by a two-thirds majority that there is not enough work to support the 2,500 consulting oil geologists. It is a group, therefore, which

[2] Ibid., p. 16.

Scientists in Society 199

may be expressing views which will become increasingly common if unemployment is widespread among scientists:

> There is criticism that some consultants are, on the whole, not very objective.
> For some, there is need to control disreputable or unethical operators.
> Six persons complained about moonlighting by major-company geologists. Many correspondents pointed out that a geologist cannot expect to make a living on purely consulting work on the basis of fees. Such fees are usually $100 a day, but it is reported that there has been undercutting to as little as $25 a day.
> To some, the rosy picture of the future need for more geologists, expressed in certain publications, is overoptimistic. One correspondent used the word "garbage."

The attitude of the geologists can be taken as a harbinger of a Guild of Scientists. How will it provide "mutual aid and protection"? Let us count the ways.

1. Limit the labor supply to eliminate unemployment.
2. Seek to increase salaries.
3. Fight for the special benefits necessary for workers in rapidly changing fields.
4. Threaten joint action if the demands of the guild are not met.

The last of these items is the crux of successful group action. Does the threat of a strike by scientists have credibility? The initial reactions I have received to this question show how set scientists are in their detached attitude. Most seem to think that the services of scientists have no immediate value to back up a strike threat. If trash men stop collecting, their negotiating strength soon becomes obvious. What is immediately threatened if scientists go on strike?

However, scientists like to solve problems, even ones that concern their well-being, and a few suggestions have now emerged. The first was advanced by a high government official. The scientists can say, "From this moment we do not guarantee the reliability of our results."

Scientists also keep discovering things of value. The announcement "We have found the means of abating jet noise but cannot seem to find our notes" might arouse the interest of management officers if they were nodding during labor negotiations. Likewise, "Some new calculations suggest that in the interests of safety, before this billion dollar reactor actually goes critical, the influence of the new element configuration ought to be simulated in the computer. Unfortunately, Dr. Jones, who is the only man experienced in this work, is so upset by the threatened strike that he fears it may take him six months to do the programming."

These matters require timing, but so does any strike threat. Trash men strike in the summer heat, airport baggage handlers just as the tourist season begins, and agricultural workers as the fruit ripens on the vine.

These "threats" sound like jokes, or are meant to, but doubtless effective tactics will be developed once a cadre of organizers put their minds to it. Consider for the moment the remarkable possibilities if scientists decide to work to rule. It is almost impossible to tell whether a scientist is actually working or merely appears to be. In fact, the moment of discovery often occurs when a scientist is not working at all, and it might well be prevented if he is kept at his desk. I have observed earlier that the problem is not one of finding enough scientists but of obtaining some scientific results. Dull and brilliant work take the same time. A worthless report contains the same number of pages as a valuable one. In any reasonable time span, who is to judge the difference other than the members of the guild? Well, of course, scientists don't do that sort of thing but, of course, they don't get fired and they aren't unemployed either. The young men who have to search long and hard for jobs may be expected to organize and fight to hold them, and they are increasingly numerous.

Granting that some muscle will appear when it is needed, how will the guild act to achieve its aims? The size of the labor force will be most important. American scientists come from three sources, namely: (1) university graduates majoring in science, (2) upward movement of scientific technicians and assistants, and (3) immigration of foreign scientists. All of these sources have been encouraged to produce in the past because of a shortage of skilled scientists. All presumably can be discouraged from producing when a surplus develops. It will be the function of the guild to control this.

The supply of new college graduates can be limited, but it will require the acceptance by colleges, and the governments that fund them, of the concept that supply should be related to demand. This may be difficult to achieve, because the trend has always been upward. The justification for more support for a department is an increase in students. It will not be an easy step for a department to refuse or discourage students on the grounds that it is illogical to spend four years learning how to be unemployed. Happily, college students are not stupid, and many will avoid majoring in a field with few jobs to offer. Regrettably, they are not very well informed about future employment. Geology majors at Franklin and Marshall College were asked, "Why did national reports of a decrease in the number of job opportunities in geology not cause you to enter another field?" and 45 percent said they had not heard of this.[3] We do not know how many high school students did hear of the job shortage and chose other subjects. We only know that about half the majors did not consider the chance of getting a job to be worth investigating. Perhaps in these days of prosperity, no white male American college graduate really worries about whether he can make a living, and thus these majors are merely typical. Perhaps, however, better information and counseling would exert more influence on selecting a major.

The guild could also exert control over the number of graduating majors. In the first place, the professors would be members and thus concerned with the security of the profession. In the second place, the guild could threaten to disqualify graduates if the faculty-student ratio was below a certain level. Finally, it could set the examinations for guild membership high enough to exclude all but the desired number of new members in the profession. This last approach would work regardless of the number of graduates pouring from colleges, and it would maximize the quality of the profession, at least with regard to the ability to pass an examination. However, it is not a very humanitarian solution to the problem of oversupply.

Upward mobility of skilled and imaginative technicians should be encouraged; a college degree is not necessary to do research. However, the guild, as opposed to the General Union, would be restrictive, and technicians would have to pass a series of examinations and presumably an apprenticeship to qualify for membership. The severity of the examinations surely would vary inversely with the number of jobs available.

The remaining source is immigration, which is now encouraged

[3] J. H. Moss, 1963, "Geology majors, are they soon to become extinct?," *Geotimes* 7, no. 6, 17–19.

on the part of skilled and educated persons having specialties in demand. Presumably the guild will inform its lobbyist to advise the appropriate members of the Congress that shortages no longer exist. It will be interesting to see what happens to the famed internationality of science if a surplus of scientists grows. Already some liberal arts colleges have been undercutting the American scientist by hiring the best foreign professor who would take the small salary available. What if such a practice actually means one unemployed native American citizen? Will the guild permit this or will the Congress? At least the underdeveloped countries and a lot of developed ones will benefit if the brain drain is stopped by barring immigration to the United States. This might well benefit American scientists as well, but it is doubtful that cutting off the cream of the foreign supply would benefit American science. On this score we may anticipate that a surplus of scientists in this country will be accompanied by a gradual decrease in the overall effectiveness of research. This may not seem very rational with regard to the interests of the nation which are protected by the lawmakers. However, the protection of the American worker against incursions of foreign labor has always been of overriding importance, and should be.

Informal guilds of scientists already exist in many fields and provide a model of what may happen. In geology there are the American Geological Institute and the American Institute of Professional Geologists, and legal registration and certification of earth scientists has begun. The AIPG is an organization dedicated solely to the professional advancement of the entire community of geologists. It is a guild in all but power.

Eligibility for membership requires:

1. graduation from an approved institution of collegiate standing with credits in any branch of geology plus a minimum of eight years of experience with a bachelor's degree, six years with a master's degree, or four years with a doctorate, plus
2. demonstration of professional competence, plus
3. unqualified endorsement of high professional and ethical standards, plus
4. full membership in an approved scientific society.

Of course, it is not compulsory for an earth scientist to join the Institute in order to earn a living, but let us just suppose for a moment that it is. At that instant the Institute would be in control of college curricula because it has to approve colleges as giving suitable training. It would completely control the size of the earth scientist labor force by its requirements of membership in approved scientific societies, professional competence, and endorsement by members of the labor force.

The founders and supporters of the Institute have no desire for such power, but in the new licensing laws we begin to see it being assembled. California has just formed a "State Board of Registration for Geologists" in the Department of Professional and Vocational Standards. Geologists who wish to practice as consultants within the state are required to be certified by the Board. The application form requires information on years of undergraduate study in the geological sciences, membership in professional organizations, names of three geologists who are "familiar with your work," and names of endorsers of "character and business integrity." It also requires a list of experience. All of these requirements seem reasonable if the Board is to do its job.

Suddenly, however, the livelihood of some earth scientists is put in jeopardy. Can a geochemist with undergraduate and graduate degrees in chemistry do consulting? The certification is merely for authorization to do consulting, but it is hard to believe that it will not come to be required by commerce and government as a condition of employment or advancement especially when a labor surplus exists. An opinion by an unregistered scientist, for example, might soon become of little value in a court of law. The registration act moreover does not include teachers among those responsible under the act, but how long will that last? The teachers will have to prepare students in the future to pass some examination for certification. Teachers will surely meet the standards they require of the students.

Viewing all these developments as inevitable, my colleagues and I have all reluctantly applied for registration and sent in checks for $40 to the Board. The imbalance between supply and demand for scientists is small at the moment. If it is allowed to continue to grow, the present requirements for registration as a scientist may seem very mild indeed. Consider, for example, one slight change in the present procedure. Professionals of all sorts — physicians, lawyers, and scientists — all say they are the only judges of the qualification of an individual to enter the profession. If qualified, a geologist can become a member of a professional society. What if the Board of Registration decides that, for simplicity, membership in the American Institute of Professional Geologists is the single requirement for registration? At one step a powerful guild is formed.

Once the size of the labor force is under control, the guild officials can turn to the salary structure with the faint hope of achieving the same wage scale as craft unions.

At present, a trade unionist may go to a trade school for a bit, then he is a paid apprentice, perhaps continuing on with some schooling, and next he is a journeyman making such handsome sums as to cause home owners to do their own repairing and painting if they can. Let us assume that the unionist graduates from high school, although that is hardly essential, and enters his trade at age eighteen. A decade later he will certainly be a journeyman, and meanwhile he will have earned a substantial amount of money, established seniority, built up social security funds toward retirement, and earned a long annual vacation.

Contrast the scientist who graduates from high school at the same age. A decade later he has a doctorate and takes his first job. Meanwhile, he not only has not made money but has spent $10,000 to $25,000 for an education. Most students work in summers and some receive fellowships, but let us consider one who is supported by his parents or by an insurance fund established to give him an education. In a general sort of way, it appears that his lifetime income is greater if he enters a trade and the education money is invested when he graduates from high school rather than if he obtains a doctorate.

Something is wrong with this system. There should be a strong and obvious incentive to encourage the best students to develop their abilities as far as they can. Let us hear no more of the mathematical genius who had to drop out of school to support his family or help in her home. The people whose education will one day be essential to society should be paid to receive it. At no level should they go through a financial filter that passes only the rich.

Many of the appropriate modes of support already exist in the form of assistantships, fellowships, and, most important, jobs. These need to be expanded into a vast complex of summer and part-time jobs in industry, government, and universities of the sort operated by Antioch College. The Guild of Scientists, which can limit the supply of full-time employees, can thereby assure openings for students (apprentices). In fact, the existence of apprentice jobs introduces stability to the employment of scientists. Each student in science should be guaranteed a part-time job at pay good enough to give him the total income of a trade apprentice. Free room and board and tuition might be equivalent to part, but only part, of such pay. If such jobs or other equivalent assistantships are not available, no more students should be admitted to colleges to major in the sciences involved.

These are only suggestions but the central point is clear. The scientists at the top are reasonably paid; those at the bottom are not paid at all. In the sciences, students are apprentices and they should be paid accordingly. Anyone who thinks students will just lounge around has no idea of the sheer hard work necessary to obtain a scientific education. The problem of other college students arises; should they also be paid? Not to get a general education, although it may be important for their well-being and that of the nation. What of other professions? Surely the same rules apply as in the sciences. We need physicians; encourage them in every way possible until a surplus is identifiable. They, of course, are already operating a guild control of the labor supply which they may not want to change. But in many professions we already have a labor surplus and educational factories pouring out more. It does not seem reasonable to suggest that students be paid and thereby encouraged to enter such professions. If they want to pay to become educated, that is their right, although those already unemployed might want to attempt to discourage them.

Salaries are important to all workers. Scientists have, in addition, special problems that would be the legitimate concern of the guild. Most of these are consequences of the rapid growth of science. In nontechnological fields, things do not change unbearably during a working lifetime. A worker can get through life without having to adjust to a job that swiftly changes. Not so the scientist. Much of his energy goes to adjustment to totally new ideas, machines, techniques, people, and places. University scientists have sabbatical leaves or at least the professors do. Moreover, the means for constant self-education are to hand. The young men presumably do not yet need to adjust, and the older ones have tenure to sustain them while they do. The older scientist in industry, government, or university, if he is not a professor, has it harder. He does not have a sabbatical year to look forward to or back upon and he does not have tenure. What will the guild do to protect him when the young men are the only ones adjusted to the changing sciences?

The guild can do any of the following: enforce job security, demand time for retraining, require provision for very early retirement. Security to continue work in a superfluous job is a charge which no society can bear. Few individuals could desire such a job. It makes far more sense for industrial costs to include depreciation for workers as well as for equipment. When the education of a scientist is worn out, the cost of reeducating him should be available — in a trust fund or otherwise. This may occur in any field after it goes through a few doubling periods, whether they are 5 years or 50. If the latter, no retraining is necessary. If the former, it may be required more than once. The charges need not be large. The broadly trained man may need only a little leisure time and expenses to attend professional meetings in an unfamiliar specialty. Those with somewhat narrower training may need afternoons off, with pay, to take courses or may attend a summer seminar for several months. The minimal scientist with relatively narrow training, or the one long specialized in a narrow field, is most in need of support. He may have to return to formal training for a year or two to learn a new specialty or refurbish a weathered education. The educational and living costs of these people are a charge on the technological system that used them and made their education obsolete. They have to be depreciated just like the equipment.

People age and are worn out just (a terrible thought) like equipment. Some will not be up to the rigors of reeducation, some will have only another doubling time ahead of them and would prefer to retire. It should be expected in rapidly changing science that this may occur at any age after a few doubling periods. A special plant may be built to manufacture some new thing. The cost of the thing includes money to tear the plant down and build a new one when the thing is replaced by yet another that is cheaper or more efficient or smaller or whatever. The cost of early retirement for the scientists who discover and develop these things should be included as well.

The universities will also have to consider what it will cost them to not offer early retirement to scientists who want it. By early I mean age forty or fifty. The departments are almost fully staffed and largely with young men. The universities thus find themselves in much the same personnel predicament as an army after a major war. How do they make openings for the new men?

Full employment, pay for students, reeducation, early retirement, support of a Department of Science, these are matters that would concern a Guild of Scientists. When enough scientists are concerned, they may organize to form the guild. It is a logical step but, I think, only a step. Almost everyone directly committed to technological work has similar problems. In the long run, and hopefully from the beginning, scientists would benefit from a league with these others in the equivalent of a broad industrial union.

THE GENERAL UNION OF TECHNOLOGISTS AND SCIENTISTS

A labor union of engineers, technicians, and scientists would have a potential membership of more than two million people who share the problems of working in a changing world. Everything that applies to a Guild of Scientists does to a general union as well — only more so. The threat of a strike is more obvious. The need for a limited labor pool, retraining, early retirement, and student pay are more acute. The marginal technician probably needs complete retraining in a new specialty more than once, and he needs pay while he is getting it. The best technician may be at a loss when the instrument or technique of his specialty becomes obsolete. When it does, he does (again a terrible thought), unless provision is made for retraining. Likewise, the engineer who designs ever better steel widgets for twenty years may simply need early retirement when they are completely replaced by fiberglass gadgets full of miniaturized circuitry. In technology, as distinguished from science, training is generally lower and specialization more intense. A small development in technology has more effects on labor.

Thus, engineers and technicians would be wise to organize just as scientists might, but why should all three organize together in a general union? The members doubtless are in different economic strata and have different political views, so what would be the advantage of the union as opposed to several guilds? The principal one would be political in the sense that the common concerns of the union members would be conveyed to politicians in the most effective way, namely, with a backing of power. The leaders of the general union will be sustained by the thought that the members are literally in direct control of almost everything that makes this nation work.

The Teamsters run the trucks, and the various service unions control essential services, and thus they are listened to when they have problems. By this stage in the technological revolution, however, almost every aspect of society depends at some level upon people with technical training. Who conceive, invent, develop, install, operate, service, and program the computers that now handle almost all banking operations in this country? Ditto stock market operations. Ditto industry bookkeeping. Ditto the auditing of the Internal Revenue Service.

Are all the quality control, flow control, automatic milling, and other advanced machinery invented, installed, and serviced by truck drivers or longshoremen? or by the technologists whose skills may become obsolete when the instrument does?

Where did the transistor come from? Likewise radar, radio, television? Whence rayon, dacron, nylon? Photography? Who finds the oil and gas and uranium that drive electrical power systems? Who devised the systems?

Technologists and scientists have common cause and much to gain by organizing. Their leaders should be speaking for them to the President, congressmen, and government and industry officials. Increasingly they will be gaining public office themselves, and their backgrounds will influence legislation and national policy. In many cases a general union might be well served by supporting their campaigns, and that in itself might encourage union members to run. A very large union, affluent as it would be with dues, would be heard. The prosperity of the United States depends in large part upon the inventiveness and scientific and technical talents of its citizens. Perhaps when these particular citizens find themselves with a labor surplus they will also find the time and the leaders to exert an influence proportional to their contribution.

Index

Abstracts, 25, 146; *Chemical Abstracts*, 14, 40, 72; growth of, 14, 40–42; *Biological Abstracts*, 40, 42; *Physics Abstracts*, 40, 50, 52, 72
Acousticians, 52
Acoustics, 12, 50–53, 61, 74
Adams, John, 65
Adams, John Quincy, 65
Aerospace industry, 4–5
Agassiz, Alexander, 93
Agassiz, Louis, 34–35, 152
Akademiia Nauk, 24, 85
American Academy of Arts and Sciences, 65
American Association for Advancement of Science, 66
American Geological Institute, 66, 202
American Geological Institute Committee on Geological Personnel, 66
American Institute of Professional Geologists, 197, 202, 204
American Journal of Botany, 46
American Journal of Science, 43, 104, 121
American Naturalist, 91
American Philosophical Society, 65
Anatomists, 62
Antioch College, 204
Arithmetic growth, 40–47, 55, 109–110
Army Map Service, 185, 186
Association of American Geologists, 66
Association of American Geologists and Naturalists, 66
Astronomers, 37
Astronomical Journal, 36
Astrophysical Journal, 30, 31
Astrophysics, nuclear, 30–31, 34, 36, 49
Atomic Energy Commission (AEC), 165, 186, 193
Authors: prolific, 87–93, 100–102

Bibliographies, 13–15, 25, 31, 130, 134–136. See also Biographical Memoirs; Science Citation Index
Bibliography and Index of Geology, 27
Bibliography and Index of Geology Exclusive of North America, 27
Bibliography of Fossil Vertebrates Exclusive of North America, 31
Bibliography of North American Geology, 27, 33–36, 40, 53–54, 62–63, 98, 109, 134
Biochemistry, 14, 47
Biographical Memoirs, 93, 97
Biological Abstracts, 40
Biologists, 14, 61
Biologists, microbiologists, 62
Biologists, molecular, 17
Biology: growth of, 25, 40, 61, 62; output and funding of, 81; doctorate trends, 164; doctoral feedback, 172
Biology, molecular, 49; DNA, 125
Biophysics, 47
Blackwelder, Eliot, 112
Botany, 46–47
Bryan, Kirk, 100, 112, 126
Budgets, 75–80, 147–149, 185–186, 187–189, 193
Bulletin of the Geological Society of America, 27, 34, 36, 45, 98, 104, 111, 121, 136, 137–138, 141, 146, 179
Bulletin of the U.S. Geological Survey, 27, 133
Bumstead, Henry, 93, 103
Bureau of Mines, 185, 186, 189
Bureau of Standards, 78
Bureau of the Budget, 76, 79, 192

Caltech (California Institute of Technology), 180

Career: promotion, 1–5, 24, 110, 147; retraining, 4–5, 182, 206; professional life span, 7, 18; in slow, average, fast subfields, 17–19; honors, 22–23, 84–85, 91–93; satisfaction in, 24, 183–184; stages of, 99–102; choice of field, 171, 183–184; early retirement, 206. *See also* Scientific fame
Census: problems of, 12–15; U.S. Census, 58–59; NSF census, 58–62; of geologists, historical, 62–66; of earth sciences, graduates, 160–163
Chamberlin, T. C., 83, 144
Chemical Abstracts, 14, 40, 72
Chemistry, 137; growth of, 25, 42, 58–61; paper output in, 40–42, 71–72, 74; transitional fields, 47–48; distribution of productivity, 109, 111; doctorate trends, 164
Chemists, 12, 14, 48, 93, 109; census of, 58–62; population and productivity, 71–72
Chemists, analytical, 61
Chemists, industrial, 52
Chemists, inorganic, 61
Chemists, organic, 61
Citation ages: average growth, 26; in fast subfields, 27–31; in slow subfields, 31–36; distribution, 36–39, 137–139
Citations: defined, 38; distribution of, to individuals, 96, 98–100; distribution of, to papers, 96–98, 100–102; ratio of, to papers, 99–100; to first and last five papers, 103, 105, 113–114; in steady state subfields, 108–113; decay time of, 111, 138; in rapid subfields, 113–119, 126; first, to a paper, 114–115; in review papers, 118; half-life of, 119; in scientific revolution, 119–125; as measure of influence, 126–128; in dormant science, 137–139
Citers of papers: in fast subfields, 115–116; relation to citee, 116; age distribution of, 116; in scientific revolution, 124
Classification systems: geological, 140–141
Colston Papers (17th Colston Symposium), 120
Committee on Professional Employment Standards, 197
Computer programming, 14–15, 62
Congress, 9, 185, 186, 190, 202
Continental drift, 83, 119, 122
Controversy in science, 142–145
Cope, E. D., 91, 92–93
Corps of Engineers, 76

Cousteau, Jacques-Yves, 84
Crick, F. H. C.: DNA, 125; Nobel Prize, 125
Curriculum and Standards Committee of the National Association of Geology Teachers, 179
Cushman, J. A., 91

Daly, R. A., 114–115, 126, 143; citation distribution, 104–106; citers of, 106–107; citation age distribution, 107–108
Darwin, Charles, 84, 143
Davis, William Morris, 93, 143
Department of Agriculture, 75, 77, 78, 192, 193
Department of Commerce, 186
Department of Interior, 77, 186
Department of Labor, 192
Department of Professional and Vocational Standards (California), 203
Department of Science: proposed, 186; need for, 192–193
Department of Science and Technology: proposed, 191–193
Department of Transportation, 192
Dietz, Robert, 120
Digest of Appropriations for the Support of the Government of the United States, 76, 79
Division of Geographic Names, 132
Doctor of Science degree, 175
Doctorates: as output of education, 159–160; census of, in earth sciences, 160–163; in sciences, 163–165; effect of depression, 165; in geological subfields, 167–171; doctoral feedback, 172
Dormant science: decay time of citations, 111, 138; bibliographies, 134–136; geology, 134–136; citations in, 137–139; jargon, 140–141; controversy in, 142–145
Doubling times: of scientists, 7, 17–19, 159; of literature, 18, 19–20, 26, 40–42; in rapid fields, 27; in slow fields, 31; of doctorates, 172–173
Dutton, Clarence E., 148–149

Earth and Planetary Science Letters, 121
Earth sciences, 22, 54, 92, 119, 160–161, 182. *See also* Geochemistry; Geology; Geomorphology; Geophysics
Earth scientists, 21, 71, 92, 146; certification of, 202–203
Ecology, 137. *See also* Environmental science

Index

211

Education: professorial influence on, 126–127; as industry, 157–160, 182; output of doctorates, 159–160, 163–165, 167–173; output of graduates, 159–163; faculty-student ratio, 166–167, 201; growth in subfields, 167–171; capacity of system, 167–175; doctoral feedback, 172; professors in, 175–178; curriculum changes, 178–180; in steady state field, 178–182; retraining, 182, 206; student support, 204–205
Einstein, Albert, 84, 102, 195
Employment, see Career; Scientists
Engineering, 163–164, 172, 173
Engineers, 3, 4, 12, 13, 58, 194; union of, 195, 207–208
Environmental science, 75, 126, 137; concern of federal agencies, 77, 186–187; federal funding of, 187–189; growth of, 187–190; formation of new agencies in, 190–191
Ewing, Maurice, 122
Executive Branch, 9, 185, 190
Executive Office of the President, 187
Exponential growth, 6

Faculty-student ratio, 166–167, 201
Federal agencies, 11, 185–186; publications of, 147–149; environmental sciences in, 186–191. See also individual agencies
Federal funding, 147–149, 185–186; ratio to GNP, 7–9; effect on growth, 75; research and development, 75–81; military budgets, 76–79; GI Bill, 165; environmental science, 186–189
Federal science, 9, 11, 154–156; funding of, 75–80, 147–149, 185–186; agencies of, 185–186; environmental science, 186–191
Franklin and Marshall College, 201
Franklin, Benjamin, 65

Geochemistry, 14, 22, 55, 138, 170, 182
Geochemistry, isotope, 111, 182
Geochemists, 146, 170, 203
Geodesy, 22
Geographers, 22
Geography, 22, 25
Geological literature: bibliographies, 31, 134–136; growth of, 40–45, 53–55; components of, 42; of state surveys, 43–44, 150–154; prolific producers, 87–93; citations in, 98–101, 104–106, 111–119, 137–139; distribution of productivity, 109–111; 19th-century productivity, 111, 134; citation decay in, 111, 138–139; scientific revolution, 119–124; editorial restrictions, 131–133; publication delay, 133, 138, 145–146; jargon, 140–141; controversy in, 142–145; of U.S. Geological Survey, 147–149
Geological Society of America, 66, 91, 120, 135, 146
Geological Society of London, 142
Geological Survey of Great Britain, 43, 150
Geological Survey of the Territories, 77
Geologists, 14, 37; salaries, 21–22; historical census of, 62–66; population and productivity, 71–72, 74; prolific producers, 87–93; fame, 87–93, 96; citations to, 98–100; engineering, 197–198; attitude toward employment, 197–199; oil, 198–199; job opportunities, 201; certification of, 202–204
Geologists, marine, 107, 118, 124
Geologists, mining, 52
Geology: growth of, 21–23, 25, 40–42, 53–55, 82; subfields identified, 54; history of societies, 64–66; history of, 82–83; scientific revolution in, 119–124; as dormant science, 134–146; classification systems, 140–141; controversy in, 142–145; number of graduates, fluctuation, 160–163; doctorate trends, 163–165, 167–171; number of professors in, 166–167; thesis subjects, 167–169; doctoral feedback, 171–172; curriculum changes in, 178–179
Geology, astrogeology, 55
Geology, economic, 54, 169, 180, 181
Geology, experimental, 55, 181
Geology, glacial, 34–36, 42, 47, 52
Geology, marine, see Marine geology
Geology, petroleum, 55
Geology, regional, 55
Geology, structural, 55, 179
Geomorphologists, 107, 109, 112
Geomorphology, 109, 170, 179, 180
Geophysicists, 22, 122, 146, 180
Geophysics, 22, 138; growth of, 27–30, 55; citation ages, 28–30; growth of doctorates, 169, 171–172; curriculum in, 180; relation of doctorates and literature, 180
Geotimes, 66
GI Bill of Rights, 165
Government agencies, see Federal agencies

Graduate Record Examination, 128
Gross National Product, 3, 7, 9
Grout, Frank, 107
Growth rates: military, 1–3; of scientific literature, 6–8, 10–11, 19, 40–47; in model subfields, 17–19; measuring of, 25; doubling times, 26–36; war, effect on, 27–28, 30–31, 33, 40–41, 52; arithmetic, 40–47; in transitional fields, 47–50; of federal science, 154–156
Guild of Scientists, 195, 197–207

Hall, James, 152–154
Harvard University, 104, 179
Heezen, Bruce, 116
Hess, Harry Hammond: sea-floor spreading, 119–120, 123–124; citation distribution, 120–124
Hydrology, 22
Hyerdahl, Thor, 84

Immediacy of science, 7, 39, 70
Industry: aerospace, 4–5; retraining, 5; employment, 70; education as, 157–160, 182–183
Institute for Scientific Information, 93
Internal Revenue Service, 208
International Geological Congress, 146
Invisible Colleges, 116
IQ, 128

Jargon, 130, 140–141
Jefferson, Thomas, 65
Joint Commission of the Congress, 187
Journal of Dairy Science, 125
Journal of Geology, 34, 104, 121
Journal of Geophysical Research, 27, 121, 146
Journal of Molecular Biology, 117, 125
Journal of Paleontology, 33
Journal of Petrology, 121
Journal of the American Oil Chemists Society, 125
Journal of the Cushman Foundation, 91
Journal of the Optical Society of America, 118
Journals: defined, 6, 36; general growth of, 6, 25; citation age distribution, 36–38; control of, 91–92; publication delay, 133, 145–146. *See also* individual journals

Keyes, C. R., 91
Krause, Dale, 123

KWIC, 15, 102

LeConte, Joseph, 178
Levorsen, A. I., 160–161
Life sciences: doctoral feedback, 172

Marine geology, 34, 49, 138; growth of, 27–28; citations in, 29–30, 113–117; sea-floor spreading, 119–124
Marine Geology, 113, 121
Marine Geology of the Pacific, 114, 115
Martino, J. P., 9–10, 166
Mathematicians, 62, 87, 174
Mathematics, 48, 61, 62, 164–165, 172
Matthew effect, 96, 98, 117, 119
Matthews, Drummond, 119
Menard, H. W.: citation age distribution, 113, 116–117, 118–119; citation distribution, 113–115, 116–117; age distribution of citers, 115–116
Merrill, G. P., 43, 64–65, 150, 153
Merton, R. K., 96
Military, 4, 29; growth rates, 1–3; budgets, 76–78; research support, 78; environmental science, 186–187
Mineralogy, 12, 140, 179
MIT (Massachusetts Institute of Technology), 176

Nader, Ralph, 185
NASA (National Aeronautics and Space Administration), 5, 186, 193
National Academy of Sciences, 9, 85, 91, 92, 147, 151, 186, 195; age of members at election to, 22; number of papers at election to, 92–93; citations to members of, 97
National Association of Geology Teachers, 179
National Commission on Marine Science, Engineering, and Resources, 187
National Register of Scientific and Technical Personnel, 66
National Science Foundation, 9, 66, 193; Federal funding, 7, 79, 148, 186; census, 58–62, 72–74
Nature, 113, 117, 120, 121, 125, 146, 184
Newton, Isaac, 53, 195
Nobel Prize, 22, 70, 85, 125
Nuclear Astrophysics: A Bibliographic Survey, 30

Oceanographers, 11, 22
Oceanography, 4, 113; federal agencies, 11; salaries, 22; cost of, 77; growth

Index

of doctorates, 170–172; mode of expansion, 172. *See also* Marine geology
Office of Education, 161
Office of Naval Research, 78
Office of Science and Technology, 185
Office of Standard Weights and Measures, 78
Optics, 12–13, 50, 52–53, 74
Ornithologists, 64, 183
Ornithology, 93

Paleoecology, 55, 182
Paleontologists, 22, 64, 92, 112, 179
Paleontologists, invertebrate, 109
Paleontologists, vertebrate, 33, 38, 65
Paleontology, 12–13, 22, 36, 55, 92, 179
Paleontology, invertebrate, 54, 109, 170, 182
Paleontology, vertebrate: growth of, 23, 31, 33–34, 35, 42, 52; bibliographies in, 31; citations in, 33–34, 47, 113–114; growth of doctorates, 171
Pan American Geologist, 42, 91
Parkinson, C. N., 149
Parkinson's Law, 8, 74, 148, 149, 154
Pauling, Linus, 102
Peter Principle, 175, 176, 177, 178
Petrologic studies: a volume to honor A. F. Buddington, 120
Petrologists, 107, 140
Petrology, 179
Petrology, igneous, 182
Physical Review Letters, 117
Physicists, 12, 14, 93, 109, 158, 167; census of, 61, 72–74; population and productivity, 72–74
Physicists, applied, 122
Physics, 12, 36, 137; growth of literature, 25, 72, 74; growth of, 40–42, 47, 61; paper output, 40–42, 47; transitional fields, 47; subfield growth, 50, 52–53, 74; output and funding, 81; distribution of productivity, 109–111; doctorate trends, 163–164
Physics, atomic, 53, 61, 164–165
Physics, molecular, 53, 61
Physics, nuclear, 50, 52, 74, 165
Physics, solid state, 50, 52, 74
Physics Abstracts, 40, 50, 52, 72
Physiologists, 12
Plant physiology, 49
Plant Physiology, 146, 147
Plate tectonics, 117
Population: growth rate, 7; of scientists, 9, 58–67, 71–75
Powell, John Wesley, 187

President's Science Advisory Committee, 9, 23, 185
Price, Derek J. de Solla, 6–7, 10, 96, 110, 116
Professional organizations: proposed, 192–193, 195, 199. *See also* individual societies
Professors, 9, 157–158, 166, 172–178
Promotion, 1–5, 24, 109, 147, 177

Research: federal funding of, 7, 75–81, 147–149, 185–186; number of scientists in, 67; state funding, 151
Research and development (applied research): federal funding of, 7, 75–79; number of scientists in, 67
Research reports, 137–138
Review papers, 38, 118, 136, 137–138
Rogers, H. D., 151
Rogers, J. B., 151
Rogers, R. E., 151
Rogers, W. B., 151
Royal Society, 24, 85

Science: defined, 6, 25; immediacy of, 7, 39, 70; transitional field growth, 47–50; subfield growth, 50–57, 167–171; funding of, 75–81, 147–149, 185–186; civilian control of, 78; mode of expansion, 126, 172–173; controversy in, 142–145; doctorates in, 159–160, 163–165, 173–175; Department of Science, 186, 192–193; Department of Science and Technology, 192–193; future of, 194–197. *See also* Federal science
Science, 113, 121, 146, 184
Science Citation Index, 39, 93, 96, 97, 125
Sciences, *See* individual fields
Scientific fame, 18–19, 22–23, 84–85, 87, 91–93
Scientific literature: growth rates of, 17–19, 25–36; war, effect on growth, 33, 40–41, 52; citation age distribution, 36–39, 137–139; classified, 52; model of development, 130; jargon, 130, 140–141; editorial restrictions, 131–133; in dormant fields, 131–146; publication delay, 133, 138, 146; of federal agencies, 147–150; of state geological surveys, 150–154; and doctorates in subfields, 181–182. *See also* Abstracts; Bibliographies; Journals; Review papers; Scientific papers; Geological literature
Scientific papers: total number of, 6;

geology output, 71, 74; chemistry output, 71–72; physics output, 72–74; acoustics output, 74; optics output, 74; federal funding of, 79–81; compulsion to write, 87, 91, 131; review process, 87, 127; prolific producers, 87–93, 100–102; number of, at election to NAS, 93; ratio of, to citations, 99–100; limit on quantity of, 101–102, 127; blizzard of, 102; publication delay, 138, 146–147
Scientific revolution: continental drift, 119, 122; sea-floor spreading, 119–124; citations in, 119–125; effect of new paradigm, 123–124; citers in, 124; DNA, 125
Scientific societies, 65–66. *See also* individual societies
Scientists: professional life span, 7, 17–19; population of, 9, 58–67, 71–74; age distribution, 17–19; salaries of, 21–22; honors, 22–23, 84–85, 91; employment distribution, 67–70; as paper producers, 70–74, 87–93; limit on number of, 74, 199, 200; as prolific producers, 87–93, 100–102; ratio of citations to papers, 99–100; limit on output, 100–102; superstars, 102; supply and demand of, 157–158; guild of, 195, 197–207; union of, 195, 207–208; attitude toward employment, 197–199; sources of, 200–202; certification of, 202–204; problems of rapid growth, 205–206; early retirement for, 206. *See also* Career; Scientific fame
Scripps Institution of Oceanography, 124
Sea-floor spreading, 119–124
Shepard, F. P., 107
Simpson, George Gaylord, 22
Smithsonian Museum, 77
Social sciences: doctoral feedback, 172
State Board of Registration for Geologists (California), 203–204
State geological surveys: literature of, 42, 43–44, 150–151; establishment of, 150–151; costs, 151–154
Steady state: defined, 3–4; glacial geology, 34–36; geology, 40–41, 53–54; physics, 40–41, 53; arithmetic growth, 40–47, 109–110; effect on education, 178–182
Stratigraphy, 55, 170, 179
Students: graduate, 19–20; as output of education, 159–163; student support, 204–205. *See also* Doctorates
Style Manual, 132

Subfields: doubling times, 17–18; models of, 17–19; age distribution of scientists, 18; effect on career, 18–23; volume of literature in, 19–20; citation ages in, 26–36; citation age distribution, 36–38; growth rates in, 50–57, 108–119, 126, 169–171; growth compared to stock market, 55; productivity of scientists in, 108–111; citations in, 108–119; doctorates in, 167–175, 181–182. *See also* individual fields
Suggestions to Authors, 131, 132

Taxonomists, 17
Teaching: future numbers required, 9; number of scientists in, 67, 70; past demand for, 158; capacity of system, 166–178; competency in, vs. research, 175–178; curriculum, 178–180
Technicians, 12, 58, 110, 157, 194; union of, 192, 195, 201, 207–208
Technology: federal funding of, 75–79; Department of Science and Technology, proposed, 192–193; Union of Scientists and Technologists, proposed, 195, 207–208; future of, 196–197
Tectonophysics, 121
The sea: ideas and observations, 120
Transitional fields, 47–50
Treasury Department, 76, 77

Union, General, of Technologists and Scientists, 192, 195, 207–208
University of California, 70, 175
University of Chicago, 176
University of South Carolina, 178
U.S. Air Force, 187, 192
U.S. Army, 2, 78, 187
U.S. Bureau of the Census, 58
U.S. Census, 12, 58–62, 71–72
U.S. Coast and Geodetic Survey, 75, 185, 187, 188–190
U.S. Geographic Board, 132
U.S. Geological Survey, 75, 151, 185, 186, 187; literature of, 42, 43–45, 148–149; publication delay, 131–133; employment, 147–148; publication costs, 148–149, 154; growth of, 154–156, 188–190
U.S. Naval Hydrographic Office, 187
U.S. Naval Observatory, 75
U.S. Naval Oceanographic Office, 185, 187
U.S. Navy, 76–77, 148, 187

Index

U.S. Signal Service, 187

Vine, Fred, 119, 123
Vocabulary, 130, 140–141

WAE (When Actually Employed), 147–148
War: World War II, 1, 27–28, 31, 41, 52, 53, 78, 158, 164, 165; World War I, 2, 40, 47; effect on growth of literature, 27–28, 30–31, 40–41, 52; Civil War, 33; Napoleonic wars, 33; War of 1812, 33; effect on education, 158, 164, 165

War Department, 76–77
Watson, J. D.: DNA, 125; Nobel Prize, 125
Weather Bureau, 185, 187, 189, 191
Wegener, Alfred, 83
Wilkins, M. H. F.: DNA, 125; Nobel Prize, 125
Wilson, J. Tuzo, 122, 123
Woods Hole Oceanographic Institution, 11, 78

Zoologists, 62
Zoology, 25, 36, 47, 93